It's in the Blood

For Pam + Mal
my very good
friends. Very
best wishes
Dave Barnah

It's in the Blood

A Life in the Meat Industry

David Barrah

TIMEBOX PRESS

First published in 2014 by
TimeBox Press
20 Chapel Road
Poole
Dorset, BH14 0JU

© David Barrah
© Typographical arrangement TimeBox Press, 2014

ISBN 978-0-9550219-2-3
British Library Cataloguing – in publication data
A catalogue record for this book is available from the British Library

All rights reserved. No part of this publication may be reproduced, stored in a retrieval system or transmitted, in any form or by any means, electronic, mechanical, photocopying, recording and/or otherwise without the prior written permission of the publishers. This book may not be lent, resold, hired out or otherwise disposed of by way of trade in any form, binding or cover other than that in which it is published without the prior written permission of the publishers.

David Barrah asserts his moral right to be identified as the author of this work.

Designed and typeset by K.DESIGN, Winscombe, Somerset
Printed in Great Britain by Hobbs, Southampton
Cover image: Keith Godwin
Cover design : Keith Godwin

CONTENTS

Acknowledgements 7

A brief history of Dave 9

Introduction 11

CHAPTER 1 **STARTING OUT** 15

CHAPTER 2 **THE LEARNING CURVE** 22

CHAPTER 3 **ABATTOIR HUMOUR** 25

CHAPTER 4 **MY COLLEAGUES AND OTHER ANIMALS**   32

CHAPTER 5 **ANTHRAX!** 37

CHAPTER 6 **THE PIGGERIES** 45

CHAPTER 7 **WASTE NOT WANT NOT** 49

CHAPTER 8 **GORDON ROAD** 56

CHAPTER 9 **LES'S MISHAPS** 62

CHAPTER 10 **MORE GORDON ROAD** 65

CHAPTER 11 **THE END OF THE ROAD** 76

CHAPTER 12	**PASTURES NEW**	86
CHAPTER 13	**DEER FARMING**	94
CHAPTER 14	**NAILSEA**	98
CHAPTER 15	**NAILSEA: THE NEW ABATTOIR**	101
CHAPTER 16	**EUROPE – BUT JUST THE START**	105
CHAPTER 17	**EUROPEAN INVASION OF NAILSEA**	114
CHAPTER 18	**LIAM FOX**	125
CHAPTER 19	**BSE CULLING**	128
CHAPTER 20	**BRIDGWATER**	134
CHAPTER 21	**MISREPRESENTATION**	142
CHAPTER 22	**ESCAPEES**	156
CHAPTER 23	**BUREAUCRACY GONE MAD**	164
CHAPTER 24	**RITUAL SLAUGHTER**	170
CHAPTER 25	**CASUALTIES**	181
CHAPTER 26	**BSE PART ONE**	204
CHAPTER 27	**BSE PART TWO**	209
CHAPTER 28	**MYCOBACTERIUM TUBERCULOSIS AKA TUBERCULAR BACILLUS OR "TB"**	220
CHAPTER 29	**TB IN BURGERS**	250
CHAPTER 30	**POISONS**	254
CHAPTER 31	**MADNESS IN COURT**	264
CHAPTER 32	**PROBLEMS FACING THE WORLD**	273
	Conclusion	279
	Epilogue	282

ACKNOWLEDGEMENTS

I DECIDED TO write a book about what I saw and became involved with at the time of the Foot and Mouth disaster in 2001. It was conceived because I saw and did so much at that tragic time and I saw much that was not seen by many. My feeling at the time was that unless these events were recorded, as things quietened down, everyone, except those who were personally involved, would soon forget all about it, as the world moved on to something else. My account of those events was well accepted by those who purchased a copy, not exactly a "best seller" but I was very pleased with how many copies were sold and also particularly by some of the feedback I received at the time and since. My now friend and publisher gave me much support at that time, as my account of the happenings of the time fitted well into his publishing business of Timebox Press, which specialises in events in history. He again encouraged me in my deliberations about my working life during fifty-two years in the meat industry for which I am very grateful. When I and just a few others have died, nobody will know how it was back then. For that reason, I got out my pen again and this book is the result.

In my first book my wife Alison was a great help, by writing out long hand what I dictated to her. My daughter Nicola transferred this manuscript onto her computer and then passed it on to Time Box Press to be made into a book. I was extremely grateful to all of them at that time for their efforts which worked out so well.

This latest document was written long hand by myself even though I now suffer from carpal tunnel syndrome in my right hand, probably as a result of so many years of gripping a knife. The legibility of my efforts left a lot to be desired and when Alison stepped in to help with spelling and punctuation she sometimes found it difficult to decipher, however she persisted and the book is now published. My daughter Nicola then took over to put it all on to computer. Sadly pressure of work hindered her efforts in this direction, but fortunately for me Nigel Guy a friend from my years in local government when working for Woodspring District Council, stepped in and very kindly put most of the second half on the book onto his computer. This involved me giving him two rings on his phone whereupon he would ring me back and set his time alarm. His phone could be in use for just under one hour without charge. He would send chapters when they had been corrected for spelling and punctuation over to Nicola's computer and she would then send them on to Time Box Press. A very long drawn out method of writing a book! However with many thanks to these people it all seemed gradually to come together and I just hope my readers will find the final document of interest so that all the messing about will have been worthwhile.

I am also very grateful for the support and help from my friends and colleagues, especially Andy Grist who helped me with my selection of photos, Colin Walters for taking the back cover photo which makes me look so much more handsome than I am sure I am (this is the second time he has achieved this having done the same for my first book) and finally Dr John Gallagher, who, although we have never met, supplied me with the photos of the tubercular badgers – I am very grateful.

A BRIEF HISTORY OF

DAVE

Dave is born	1943
Helps at his uncle Billy Nash's shop and abattoir	1950
Dave goes to school at Clifton College, Bristol and leaves after his 'O' levels	1960
Farm experience	1960–61
Butcher's shop (John Clark)	1961–63
Meat Inspector's Assistant (Bristol City Council)	1964
Qualified Meat Inspector	1965
Works at Hotwells Abattoir and Spear's Abattoir	1965–66
Works at Gordon Road Abattoir and Spear's Abattoir (Spear's Abattoir closes in the 1970's)	1966–81
Dave promoted to Senior Meat Inspector for Bristol City Council	1977–81
Woodspring District Council – Meat Inspector	1981–95
Old Nailsea Abattoir	1981–88
Weston-super-Mare Abattoir, Blagdon Abattoir, Clevedon Abattoir and the deer farm all within	1981–92

New Nailsea Abattoir	1988–95
Meat Hygiene Service formed and Dave works as a casual inspector	1995
Dave leaves Nailsea	1999
Cull abattoir working on the BSE cull	1999–2001
Foot and Mouth and Farm Animal Welfare scheme	2001
Food Standards Agency created	2001
Bridgwater abattoir working on the BSE cull	2002–6
Nailsea abattoir	2006 – present
Bristol University, School of Veterinary Sciences, visiting lecturer	early 1980's – present

INTRODUCTION

IT'S IN THE BLOOD

IT ALL STARTED when I was about seven years old. Whilst on summer holidays in Pembrokeshire I spent much time in the company of an uncle, Billy Nash, who owned a butcher's shop in Neyland. As was common in those days, the animals were killed by Billy and his son in a small slaughterhouse behind the shop. These animals would have been purchased from local farmers or bought at auction in the local mart or cattle market. Many of the incidents I witnessed in those far off days as an impressionable youngster would today cause the army of bureaucrats who control the meat industry to throw up their hands in horror.

Those early experiences had a great influence on the direction of my working life, spent in the abattoirs in Bristol and the nearby county of Somerset. However, this was not necessarily how it was meant to be! On leaving school I was set for a career in farming, and had done the best part of my year's practical training prior to going to agricultural college when I was struck down by a mysterious illness. Medical advice suggested that I should spend some time indoors, away from the rigours of the weather, so I started working on a temporary basis for a family friend and patient of my father's, Jack Clark, who owned a butcher's shop in Henleaze in Bristol. At this point I should say that my father was a successful physiotherapist, with a private practice in Bristol, and would most likely have liked me to have followed in his footsteps. But human anatomy was of no interest to me; whereas animals and their anatomy were of riveting interest.

Sadly, I did not apply myself to academic study at school, which would have equipped me for a career in veterinary science. I was far happier in my own little laboratory at home, boiling up the bodies of foxes, badgers and anything else I could find to extract and rebuild the skeletons. I accumulated a large collection of skulls and other body parts.

At the time we lived in a large four storey house in Bristol adjoining the Durdham Downs, and it was my habit to store many of these bodies (in various stages of decomposition) on a flat leaded area of the roof which was accessed by a trapdoor above my laboratory.

At around this time radio crystal sets were the "must-have" gadget amongst my peers. My older readers will know that these crystal sets needed an extensive length of copper wire as an aerial to receive the signal. On top of the roof seemed the ideal place for my aerial, and so I stretched the wire to all four corners, eventually anchoring it to the television aerial. A further wire was draped down the front of the house and into my bedroom window. Reception for the radio was fantastic, but unfortunately at almost exactly the same time the picture on our television started to become very fuzzy.

One day, while I was away at school, a repairman was summoned to deal with the problem. His conclusion was that there was a problem with the aerial which would necessitate him accessing the roof through the trapdoor! He came back down much faster than he went up, telling my mother that the roof was covered in dead bodies and he was not prepared to return to examine the aerial. My father was called from his surgery to ascertain the extent of the problem. Putting his head through the trapdoor he knew immediately that

I was involved. He did manage to calm down the repairman who, having imbibed a reviving cup of tea, was persuaded to return to the roof. He then discovered my amateur attempts to collect the airwaves and the matter was easily sorted. Needless to say, on my return from school I was in double trouble and my future listening was somewhat curtailed. However, I was allowed to keep my storage area, my father being perceptive enough to see the passion I had for my subject.

The temporary job in the butcher's shop lasted for two years, and I never returned to farming. I had arrived at the stage where the natural progression would be my own shop, but instead fate took a hand. Another patient of my father's, George Whone, was a senior lecturer at Bristol Technical College, and he suggested I might be interested in a career in public health. I looked into this, but decided that the only aspect of public health that was of interest was the inspection of food animals at the point of slaughter. I therefore declined the invitation to train as a public health inspector (now called an Environmental Health Officer), and – in a decision that changed the course of my life – awaited the first course to be run in Bristol for specialist meat inspectors.

CHAPTER I

STARTING OUT

WHILE AWAITING THE start of a new course for specialist meat inspectors in Bristol I applied for a post with Bristol Corporation as a meat inspector's assistant. In those days the job was done predominantly by public health inspectors, but often this was only part of their duties. Abattoirs were already starting to get bigger, with the closure of many small family businesses. This in turn required the services of full-time meat inspectors, and the Bristol course was set up to satisfy this demand.

In 1963 I was lucky enough to be accepted, and was one of two new meat inspector's assistants to be appointed, subject to a medical examination at the central health clinic. I well remember as a fairly shy 20-year-old male that an antenatal clinic was in progress and, having been presented with a specimen jar, having to run the gauntlet of a long corridor with very rotund ladies seated all along one side to a toilet situated at the far end. After some delay, owing to my inability to produce a specimen (even after having flushed the toilet two or three times to see if running water might trigger a response – all of which I am sure could be heard by those in the corridor),

I had to return along the same route, clutching the specimen jar with its minuscule offering in the bottom, under the steady gaze of those seated. I looked straight ahead with face scarlet and tried to ignore the muffled laughter as I progressed.

Fortunately, I was deemed to be in rude health and the post was confirmed.

The post involved rotating between the three abattoirs which operated within the city limits at that time. The first was a Bristol Corporation-run abattoir. For some years after the Second World War it was a legal requirement for local authorities to supply slaughtering facilities for the local butchers to kill their stock, as many private premises had been shut during the war as a result of rationing and the close control of meat supplies. The second was a bacon factory named Spears, which killed about 1,000 pigs a week over three days. Originally a Bath firm, with factories in Bath, Bristol and Redruth in Cornwall, their pork and bacon products were highly regarded in the locality.

The third was called Mutual Meat Traders. Latterly run by a co-operative of Bristol butchers, the abattoir was located at the Cumberland basin, part of the then functioning Bristol Docks, within sight of the dry dock where Brunel's *SS Great Britain* now lies. This particular abattoir was originally built to slaughter imported Canadian cattle, and was positioned in such a way that the cattle boats could tie up alongside and discharge their cargo. These cattle were then tied up in rows in the "lairage", which is an area to contain animals prior to slaughter. The capacity was allegedly 900 cattle, surely making it one of the largest facilities in the country at that time. There was an auction ring so that the cattle could be auctioned off to the Bristol butchers. The problem was that there was a restriction on the number of days that the cattle could remain alive after entering this country. Exploiting this fact, the Bristol butchers would only bid for the cattle on the last day of their lives, which resulted in rock bottom prices and the eventual closure of the abattoir.

After a time another trade emerged, importing Irish cattle which could be unloaded in a similar manner on the quayside. The abattoir was again closed during the wartime period of restriction, reopening – as already mentioned – as a co-operative, killing not only their own meat but also meat for wholesale purposes and export to Europe. The latter was flown to its destination from Bristol Airport.

My fellow appointee, Les Worden, had been working in a managerial capacity at the Mutual Meat Traders' abattoir, and it was considered by our new employer, Bristol Corporation, not appropriate for him to work in an official capacity on the premises he had so recently left. This decision restricted his work to two premises, whereas I could fluctuate between all three.

Now began a period of intensive learning. Les and I were invited to attend the meat sessions of the public health inspector's course, both theory and practice. We were very lucky to work with older and experienced officers who taught us so, so much on a day-to-day basis. Sadly, this is seldom the case today. Students and young inspectors frequently tell me there are comparatively few inspectors with long experience still working. So many resigned when the Government in its wisdom set up yet another of their agencies to run the meat inspection service in this country, and I think we are all fully aware of just how inefficient and overburdened with bureaucracy these agencies have been. Millions, if not billions, of taxpayers' money has been squandered on these ill-advised schemes. In the case of meat inspection, prior to the advent of the Meat Hygiene Service in 1995 there was a perfectly efficient system, with local authorities being responsible for delivering the service.

My first placement was at the very old abattoir situated at Bristol Docks, then functioning well below its potential capacity, killing perhaps 150 cattle, 1,000 lambs and 200 pigs per week. Compared to the vast numbers which can be killed today on fast-moving lines, this is a relatively small number, and could probably be dealt with in some plants in a couple of hours. However, as my previous experience was confined to my uncle's kill of maybe two cattle, three or four pigs and maybe 10 lambs per week, the size of this operation was initially overwhelming; also, slaughtering procedures had changed somewhat since I was a child. The small animals – sheep, pigs and calves – were now electrocuted and hoisted to have their throats cut to allow the blood to pump out, whereas hitherto when helping my uncle the sheep would be shot whilst held in the sitting position, then laid on their sides with their head in a gulley running the length of the slaughterhouse. Their throats would then be cut and the blood would flow to the nearest drain. Calves were not restrained because due to their naivety they would not try to escape, and would be shot in a standing position and bled in a similar manner.

The pigs were always killed last, and would be released to walk around the slaughterhouse munching on extraneous scraps of tissue from other species slaughtered earlier in the day. While so engaged, they would be stalked and shot with a captive bolt pistol. If you were very quick after the shooting there was always a moment of rigidity during which the animal could be stuck (bled). By pulling back on one front leg and depressing your boot on the abdomen with a pumping action it was possible to aid the bleeding process. If you missed this moment of inactivity the pig would start reflex kicking, propelling itself all around the area. You then had to wait for the animal to again relax as it started to die before you could stick it. However, there was always a danger of a condition known as "blood splashing", which is a rupturing of the small capillaries in the muscle making the flesh look as though it has measles, which can spoil its saleability. The capillaries rupture as a result of the shock of the shooting but this can also happen due to the shock of electrocution. The crucial element is the lapse of time between the stun and the stick.

Cattle were lassoed with a rope and slip ring which had previously been threaded through a metal ring in the floor. Three or four men would then tug on the rope until the animal had been pulled into the slaughterhouse and its head pulled down to the ring, whereupon it would be shot. It was not so many years before these early memories of mine that at this point a poleaxe would have been used, an implement not dissimilar to an ice axe on one side with a blade on the other side similar to a wood axe for chopping off the horns. In those days a whippy piece of cane would be inserted through the hole in the head, passing through the animal's brain and down the spinal column, smashing the central nervous system as it went: a process known as "pithing". This was done to stop the reflex thrashing of the front legs, primarily for the safety of the man who was going to cut the animal's throat, but also because it had the beneficial effect of hastening the animal's death.

Sheep and calves would have their throats cut, and often the head would be removed at the same time, achieving the same result as pithing by severing the spinal cord between the skull and the first vertebra.

The captive bolt pistol, a weapon not dissimilar to a Hilti gun for firing nails into concrete, was designed to fire a hollow bolt through the skin and skull and into the brain of animals. The bolt was contained within the gun by recoil pads, and having travelled the designated distance, returned into the gun ready for the next charge to be inserted. The advent of this weapon paved the way for a considerable improvement in the welfare of animals at the point of slaughter. When I started work there were other innovations such as the restraining of cattle in a large steel crate known as a stunning box. This was designed in such a way that the front of the box would collapse and the animal would be ejected onto the floor of the slaughterhouse where it would be dressed. Another improvement was that, after stunning, sheep, pigs and calves would be hoisted by a leg shackle onto a bar known as the bleeding rail and bled into a catchment trough.

Since those days we have moved on to electrocution and the gassing of pigs with carbon dioxide, and the only animals to legally just have their throats cut without any anaesthetic in this country are those selected for religious slaughter, be it Halal for the consumption of the Muslim population or Kosher for the consumption of the Jewish population.

One wonders when the law of the land states quite clearly that a licensed slaughterman should render an animal insensible to pain prior to slaughter, and he is carefully adjudged as to his competence to achieve this before acquiring a licence, why an exception on the grounds of religion should be made. Anyone who does not have a religious dispensation is liable to a fine or even imprisonment and revocation of his slaughtering licence. I am strongly of the opinion that the laws of this country should apply equally to all citizens, regardless of race or religion. I am sure that most people in this country are blissfully unaware of the unnecessary suffering visited upon animals which have their throats cut without anaesthetic in this way.

The original reason for the so-called religious slaughter was to stop people eating carrion; in fact it could be said that this practice was the very earliest meat inspection. The idea was that one should see the blood flow from the animal to prove that it was alive prior to being prepared for being eaten. If the animal was dead the chances were it had died of something horrible, possibly a disease which could be transmitted to man, especially if it was not properly cooked. The most serious of these would probably be anthrax, which is a certain killer. So the reasons behind the enshrining of these rules in religious belief were very commendable at the time. Unfortunately, they fail to take account of our ability now to render an animal insensible to pain prior to cutting its throat.

To return now to my placements, I was kitted out with rubber boots that had leather soles and steel studs which, when walking on the concrete floor, made me sound like a carthorse. I wore a dentist-type white coat which buttoned up the side and across the shoulder with removable buttons. One of my early duties was to extract these from all the inspectors' coats before consigning them to the laundry basket. Everybody wore a fresh coat every day, sometimes two. On my head I wore a round pork-pie type white hat with no peak and, to complete the ensemble, a long rubber apron which protected my body from high chest to ankles. I now resembled something akin to an ice cream salesman! Later, when I was allowed to use a knife, I wore a belt around my waist which supported on one side a scabbard which held two or three knives, and on the other, suspended on a piece of chain, a knife-sharpening steel.

My initial duties were very mundane: dealing with the laundry, recording the rejection of various organs and carcasses by the inspectors on a daily basis, making the tea and cooking the breakfast in the meat inspector's office, which was barely big enough to seat two inspectors and one assistant, let alone accommodate a cooker. However, necessity being the mother of invention, the two-bar electric fire – which was the only method of heating what was in winter a bitterly cold room – when put on its back and supported on two chairs, doubled as a cooker. A frying pan would be balanced on the top and breakfast, having been procured from one or other of the meat salesmen, in the form of some liver, steak or some chops or sweetbreads, would be produced by me to coincide with the half hour break at nine or ten o'clock. This always made a hearty start to our days.

CHAPTER 2

LEARNING CURVE

AFTER SOME MONTHS of experience gained by watching and assisting the qualified inspectors, complemented by private study and sitting in on the meat inspection aspects of the public health course, eventually the first specialist meat inspector's course to be run in Bristol commenced, and Les Worden and I were duly enrolled on it.

We were very ably taught by two Bristol University vets, a Mr Wilson and a Mr Ashdown. Some of our sessions took place at the Bristol municipal abattoir which had a detained room for containing carcasses and offals requiring further inspection. In the middle of this room there was a large marble mortuary slab on which the resident inspectors would assemble various diseased organs, and suspended from overhead rails would be rejected carcasses collected over a couple of days prior to our lessons. On a Wednesday evening all 20 or so of us would solemnly troop in and stand around the aforementioned slab awaiting instruction on the variety of problems exhibited in the morbid specimens in front of us.

For those of my readers who may not be aware what a meat inspector's job was all about I shall explain. The job itself was very simple: acting as a buffer between the owner and supplier of meat and offals and the consumer thereof. This involved seeing that the animals were well treated up to and at the point of slaughter, and the examination of all parts of every animal killed to adjudge its fitness for human consumption; that the dressing process was done in a hygienic manner, and that the premises in which it was done were kept clean and well maintained. That was then. The world has since moved from the sublime to the ridiculous, and much of this book will explain how I have come to this conclusion.

I was in my element, and quickly became fascinated with the variety of both parasitic and pathogenic disease problems displayed before us on a weekly basis. Forty five years later, the same enthusiasm and fascination remain, although now my pleasure is to show others what I have spent my life looking at, and to see their interest kindled in the same way.

We also had to learn all about the relevant legislation which empowered us to eventually do our job. When qualified we would be authorised as officers of the local authority to seize meat, offal and meat products which we considered to be unfit for human consumption. The owners of said items had the right to challenge our judgements before a magistrate, and only the magistrate had the power of condemnation. This process could take time, and so to facilitate the smooth running of the system the abattoir owners, when having their plants relicensed on an annual basis, would sign a form to the effect that they would voluntarily surrender any carcass or offal or meat product if deemed to be unfit to the meat inspectors for destruction, a system which in most cases worked extremely well.

Eventually the time came for our examination run by the Royal Society of Health. This was in three parts: two written papers and a practical test of morbid specimens, both carcasses and offals. I vividly remember my practical test which was on a Saturday morning. Each student was questioned for 15 minutes with one vet and a further 15 minutes with another. I completed my first session and was about half way through the second when the vet told me to stand down until the time had elapsed. I was very concerned that I had made a dreadful mistake, but later it transpired that I had more than satisfied him as to my competence in less than the required time. His estimation of my ability, plus my theoretical and legal knowledge, put me through with a good pass mark and in due course I received my certificate from the Royal Society of Health. This meant that I was now appointed as a fully fledged meat inspector: my career in the meat industry could begin in earnest.

CHAPTER 3

ABATTOIR HUMOUR

INITIALLY I WAS based at the abattoir by the docks at Hotwells, but on three mornings a week, in company with the then senior meat inspector, I worked at Spears bacon factory. Early starts are the order of the day in the meat trade, and as I had a journey of about 10 miles, it would necessitate getting up at about five am. At the Hotwells plant they worked from six am until a late lunch, with a stop for breakfast and maybe a tea break along the way. This left plenty of afternoon for other activities, which included meeting my then girlfriend (now wife) from school where she was in the Sixth Form.

As a very young inspector it was inevitable that I should be wound up as frequently as possible for the amusement of the abattoir staff. This was always done in a friendly but nevertheless aggravating way. They knew that I would be keen to prove my worth, but that in doing so I would lay myself open to their teasing.

One incident which comes to mind, and which was played out over three days, involved the wearing of washable head coverings by all abattoir staff. When starting work one morning, one of the slaughtermen was not wearing his hat, which in itself was a small enough transgression at the time. Not wishing to have a confrontation, I waited to see if the problem would rectify itself. However, as time passed it became evident that I would have to intervene. The man concerned, who subsequently became a good friend, was about my own age. I approached him somewhat tentatively, quoting chapter and verse of the legislation so newly absorbed, and asked him if he would mind putting on a hat. He said that he was unable to comply with my request because he suffered from bad headaches and his doctor had given him a note to the effect that he must not wear a hat. All the surrounding faces were absolutely deadpan: not even the flicker of a smile. I decided that discretion was the better part of valour, and I could not countermand a doctor's decree even to comply with the law.

I retired to the office to pore over the legislation to see if there was anything in the way of a dispensation that I had missed. Finding no such solution I returned to the fray to ask to have sight of the doctor's certificate, which I was reliably informed had been left at home but could be produced the next day. All I could do was await this event, so we continued as we were for the rest of the day.

The next morning there was no sign of either hat or certificate. Further confrontation revealed that it had sadly been forgotten, but was promised faithfully for the following day. The third morning arrived and still no hat and no certificate were to be seen, but this day the senior meat inspector, Les Mawditt, appeared. On entering the slaughterhouse he enquired as to why that man was not wearing a hat. I had to lamely explain that there was a very good reason for this misdemeanour, taking the form of an as yet unseen doctor's certificate. Les nearly exploded. "Certificate?" he said "What bloody certificate? Tell him to put his bloody hat on straightaway!" His raised voice was heard by the slaughtering gang who had to put down their knives while they fell about laughing, the miscreant magically producing his hat from underneath his apron and popping it on his head.

The humour in abattoirs, by virtue of what we do, is particularly black. It is a psychological release from something which most people outside the trade find distasteful; and comments and events which would make most people cringe we find hilariously funny. Having said that, I have found over the years that there are some really nice people working in this industry, and only occasionally does one encounter the psychopath.

Some of the old men I met in those early days were blessed with such names as Fred Primrose, Ted Semmery, Tom "Porky" Wintle, Bob Allsworth, Jack Uphill, Mike Rowley, Fred and Arthur Williams, Bill Davey, Pete Worle, Les "Cocky" Britton, Bill Green, and one whose real name I never knew, who was universally known as "Deafy" for obvious reasons. Their attire in those days would again make the modern hygienists throw up their hands in horror. They mostly had leather studded boots, some wore leather gaiters and some cut off the sole of rubber wellingtons and pulled the remainder down over their leather boots. Some had pieces of hessian sacking over their trousers tied above and below their knees, with which they gripped a sheep's foot when skinning the leg. Many wore dark striped waistcoats, often with a gold timepiece in the pocket on one side and a snuff box in the other. They wore linen striped shirts of the type which had detachable collars on studs, but never ever wore the collars. The timepiece would be produced at intervals to see how the day was progressing. The snuff box, which contained powdered tobacco, was surreptitiously used frequently throughout the day.

The use of tobacco in any form was forbidden in the slaughterhouse, so I had to keep my eyes about me at all times as the taking of a pinch of snuff could only take a couple of seconds, whereas the prolonged smoking of a cigarette gave the inspector more time to catch the culprit. Another particularly unpleasant habit was the chewing of something called "twist" which was tobacco formed into a liquorice-type consistency for chewing on and ultimately spitting out. Spitting in the abattoir was also forbidden, not surprisingly.

The slaughtermen's mess room, to which they retired at break times, was in fact a little room tacked onto the end of the lairage where the animals were kept. By the way of amenities there was a Tortoise stove to heat the room during winter, which could often be bitter. I well remember one winter that the water in the Cumberland basin, an annex to the Bristol docks, was frozen to a depth of a foot, and the Port Authority employed a tugboat to drive up on top of the ice and use its weight to smash the ice beneath, in order to allow the lock gates to be used for shipping. That was a cold winter.

To return to the mess room, there was a sink with an open drain beneath. Each man had his own chair in the same position every day, and each man would have his own flask of tea, which in those days, long before teabags, also contained tea leaves. Having quaffed the beverage, the residual tea leaves at the bottom of the cup would be propelled in the direction of the open drain. Over time the tea leaves which did not reach the drain would stick to the floor, and eventually a ridge would form in a line from each chair to the drain, like the rays of the sun, something that had to be seen to be believed.

All this sounds horrendous to a modern ear, but these same mostly older guys took tremendous pride in their ability to produce a clean and perfectly dressed carcass. Indeed, I remember that at the end of their hard working day, instead of rushing off home, they would wash themselves and stand back to admire and criticise their own handiwork. In fact, each individual was able to pick out the particular animals they had flayed by the thin lines left on the sides of the carcasses. These were made by their scimitar-shaped siding knives which slide gently between the skin and the subcutaneous tissues. The angle and the number of lines varied from man to man.

The skills of these men were again apparent when it came to keeping the often faecally contaminated hide away from the meat. If, as happened from time to time, small amounts of faecal material, dried earth or hair or wool should appear on the carcass, they would employ a wiping cloth, which was in fact the muslin wrapper which came around the New Zealand lambs imported extensively into the country. Each man had his quota of cloths which would be kept in a bucket of near boiling water and wrung out to nearly dry prior to use. The texture of the muslin was exactly right to gather any extraneous material which might be adhering to the carcass, and leave it absolutely dry. Each night the cloths would be put in a bucket of bleach water to cleanse and bleach them ready for the next day's work. In today's climate the very thought of using a wiping cloth in this way would appal the powers-that-be, but more of this later.

Inevitably, where a commodity is being produced, a certain amount of pilfering takes place. Next door to this abattoir was a timber yard. At breakfast time one of the workers from the timber yard would appear, and orders would be taken for particular pieces of wood or moulding, and at lunchtime pieces of meat and offal would be passed over the dividing wall and the pieces of ordered timber, all cut to size, would come back the other way. Having led a very sheltered existence up to this point, I could not believe that this sort of thing went on; however, when one considers for example the MPs' expenses scandal, this was very small beer.

The ingenuity shown by some of these guys to escape detection was unbelievable. One old man had a leather belt around his waist with meat hooks through it on which he would hang whatever he had managed to spirit away during the day. He would then cover the whole of his person with an old brown overcoat and trundle off to the bus stop, probably with an ox liver hanging between his legs and various kidneys and other tasty items hanging from his waist. He would walk with a rolling, bow-legged gait: nobody sitting next to him on the bus would imagine he was harbouring a small butcher's shop!

This particular abattoir was laid out in such a way that the men would work in teams of three or four, and each team would have their allotted quota of animals for the day. The stockman would have to take great care in dividing up each batch of animals between the men, as there would be much consternation if one team had larger cattle than another, or if there was an unequal distribution of uncastrated ram lambs, which are much harder to skin than a female or a castrate.

One of the older slaughtermen, Arthur Williams, was also a good engineer, and it was he who designed the first oscillating saw for cutting cattle in half, hitherto cattle having been chopped in half with a large cleaver or sawn down by hand. Arthur used an old motorbike engine, and his idea was later taken up by some people who improved and developed it, forming a company on the back of it called Bristol Saw Chine Company which later became known as Bristol Abattoir Equipment Company. I don't know if poor old Arthur was ever rewarded for his legacy to the meat industry. Since then, the world has moved on again, and a continuous band saw is used which is very fast and efficient.

In all abattoirs, provision has to be made by management for office accommodation for the inspectorial staff, but the office in this abattoir was not much more than a large broom cupboard. It combined its office function with being a clean and dirty laundry store, somewhere to keep our civilian clothes and, as already mentioned, somewhere we could cook and consume our meals at break times. The ill-fitting door on this office opened directly onto a yard on the far side of which, probably 20 feet away, was a bay which had been built to contain the stomach contents and clotted blood from the animals slaughtered there. This stinking pile was rarely removed, and inevitably during the warmer months of the year flies would lay their eggs upon it. Given a little time and a few warm days the pile would become a seething mass of maggots. Given a little more time the maggots would mature to a stage where they needed to pupate, and I vividly remember one particular day arriving in the morning to find the intervening yard between the pile and the office white with maggots. They were all moving purposefully in one direction which, unfortunately, was towards the door of the meat inspector's office. Negotiating the small entrance step they had wriggled their way under the door and entirely covered the floor area, as well as being behind all the furniture within.

I borrowed a sweeping brush and proceeded to try to repel this living tidal wave at least as far as the yard. However, as fast as I swept they turned around and together with their followers resumed their determined march. All I could think to do was to put an arc of neat Jeyes fluid around the doorway, which killed the first to arrive; but there were so many that their comrades wriggled over the bodies and continued on their way. My next course of action was to use a shovel to fill a bucket and then tip them into the docks, which was not far away. Unfortunately, despite my best efforts, after a while we had to endure fly hatches over several days from those maggots which had managed to crawl into crevices out of my reach.

On another memorable occasion we had rejected a couple of cow carcasses which were destined to go to the bone yard for rendering. For some reason the lorry which usually carried the bins of rejected and unusable material was out of commission and the bone yard sent a small flatbed lorry instead. The carcasses were cut in quarters and deposited on the lorry. As there was, and still is, a legal obligation to stain rejected bodies to stop them getting back into the food chain, it was decided that these quarters should be stained where they lay.

Normally a black stain in powder form would be mixed up in a watering can and sprinkled over the offending material. This day, however, our stock of the dye was considerably depleted and there was not enough left to do the job adequately. Just then the senior meat inspector arrived and decreed that our staining was not sufficient; when told we had run out of the black dye he looked amongst our stocks of marking inks etc and found a jar of orange powder marked "Drain Stain". "I know," he said "We will use some of this instead" and without further ado scattered the entire contents of the jar all over the carcasses, blissfully unaware that even a tiny spoonful of this substance can go a very long way. Just as he did so we had a cloud burst which caused the orange powder to turn a bright green.

The driver of the lorry, oblivious as to what was happening on his truck, pulled out of the abattoir entrance and drove off down the main road, a wave of liquid staining the road green as far as the eye could see!

CHAPTER 4

MY COLLEAGUES AND OTHER ANIMALS

AVONMOUTH DOCKS WERE not very far away, just a short trip down the Avon Gorge and under the famous Clifton suspension bridge. These docks could accommodate much larger ships than the city docks and were often crewed by Asian Muslim sailors.

When restocking these ships halal meat was required and these animals had to be slaughtered by a Muslim. To accommodate this situation, a ship's chandlers firm, Harding Brothers, would collect a couple of these Muslim sailors and come to the abattoir to kill the required sheep. As those in the business will know, the vast majority of halal sheep killed in this country are old worn-out ewes. Because of their age, and often infirmity, many offals and a disproportionate number of carcasses are rejected. These ewes would be specially acquired to satisfy this particular market. As is well known, relations between the Muslim and Jewish communities are not good. One only has to look at the Middle East today, and things were no different back then. In those days, my hair was extremely dark, and I have worn a beard all my working life. In addition, as I have already mentioned, in the abattoir I wore a white round pork pie type hat. As a result of this general appearance I obviously gave the stereotypical impression of being of Jewish descent.

On one particular day two of these Muslim sailors were watching me inspect the carcasses and offals we had just killed for them to restock their ship. As I remember it, this was a particularly poor bunch of sheep, and virtually all the offal and some of the carcasses were rejected. These sailors were obviously totally unaware there are such people as meat inspectors, and during the inspection process an occasional glance in their direction on my part told me that they took my activities as a religious affront. This culminated in one of them nudging the other and pointing an accusing finger in my direction, whilst shouting "Rabbi, Rabbi"!

Most of the slaughtermen had entrepreneurial tendencies, and nearly all had second jobs as a result of the usually early finish. One in particular, Belgian in origin and Gus by name, was the gutman, whose job involved stripping the intestines away from the retaining abdominal fat, emptying the gut and cleaning it for use, in the case of sheep as chipolata sausage skin, in the case of pigs as large sausage skin. Sometimes the latter would be plaited and made into chitterlings, and in the case of cattle the gut was made into gut rope apparently often used as glider tow rope. It also was used when thinned down and dried to make surgical sutures and strings for tennis racquets etc, the material commonly known as "catgut". This was in the days before synthetic materials had taken over so many uses.

One of Gus's second jobs was the running of his own maggot farm, which took place at the back of the abattoir. This involved the filling of 50-gallon metal drums with pig hair, which is removed from pigs during the scalding process. This contains not only hair but several layers of skin, and was a particularly favourite site on which the obnoxious blowflies laid their eggs. Millions of the resultant maggots, when grown to sufficient size, would be harvested by this enterprising man, who would then supply the fishing tackle shops in Bristol for financial reward, but the vast majority were used to supplement the diet of his turkeys. This was in the days before the carefully balanced rations which are available today, and was recycling in the true sense of the word.

Although in the modern world some people may find this sort of basic behaviour repugnant, it was a perfectly natural procedure, because in the wild birds will pick over the rotting remains of some deceased animal or bird to consume the maggots. In fact many game keepers back then and before would deliberately hang in their rearing pens carcasses of vermin that they had killed, from foxes through to magpies, in order for the resultant maggots to fall to earth and be consumed by the young pheasants.

This man had obviously lived through hard times in his home country, and was able to make use of anything and everything that came his way. It has to be said that I think that times like those will come again as a result of what the political class has done to us and the frivolous throw-away society which we have become. The difference between then and now is that people would not know how to survive the hardships that were prevalent in yesteryear.

Fred Williams set up a car repair business in his spare time and would look after the vehicles of most of the abattoir employees. Bob Allsworth spent his evenings as a bouncer, which didn't leave much time for sleep. I shall never forget the day when he failed to appear for work one morning. Many were the observations as to where he may have laid his head the previous night, as he revelled in the nickname "the Wiltshire ram". At about nine in the morning, when Bob still had not appeared, the manager, John Tanner, called us all together and broke the news that this very popular guy had been killed the night before in a head-on collision with a police car while riding his Vespa scooter. The feeling of shock and sorrow amongst all of us was very dramatic, and work ceased for some time; men sat shaking their heads in disbelief that this very strong vibrant man was suddenly no more. I do not remember, when in subsequent years various people have died along the way, such a universal feeling of devastation and mourning.

Bill Green was a semi-professional rabbit catcher, and with his ferrets and nets he would scour the Mendip hills, selling the resultant rabbits in the old Winford market, usually for 10 shillings, but sometimes for as much as one pound. When you consider that rabbits caught in the same way today fetch only one pound or so wholesale, obviously inflation has passed the rabbits by!

Other men stayed within their own field of expertise for extra income. Pete Worle, who was only a few years older than me, became a local legend in his working lifetime for his incredible speed and skill, and he remains a good friend to this day. He would go down into Somerset to several small slaughterhouses, mostly behind butchers' shops, where he would kill and dress their weekly requirements.

Jack Uphill would travel to Bishop Sutton, where at an abattoir on the edge of Chew Valley Lake he would bone out meat for their local shops. This abattoir later specialised in the slaughter of horses and became one of the biggest in the country doing this emotive work. I feel privileged to have worked with men of such great character, and with such diverse interests.

Not long before it finally closed, the abattoir acquired a contract for exporting lambs to France and Belgium, to be flown out from Bristol Airport, which at that time was little more than a couple of buildings and a windsock. These lambs had to be dressed in a particular way, where one hind leg had to be passed behind the Achilles tendon of the other in order to be hung on a single hook like a legged rabbit, rather than on a gambrel. As a result of this new trade a veterinary presence was required for the first time because Europe decreed that only veterinary inspected animals could be exported. Later on, when Britain went into the "Common Market", the veterinary stranglehold on the meat industry really tightened.

To satisfy this requirement a local vet who was already retained as a consultant by Bristol City Council on an ad hoc basis, and who had his own successful practice, together with occasional appearances on television, would arrive to do an ante-mortem inspection on the sheep destined for export. I would point out that, at that time, my ability to ante-mortem animals for home consumption was considered perfectly acceptable. He would then acquire an office chair which he would place in the middle of the slaughter hall, and on which he would sit and happily smoke several cigars while the killing and dressing took place. This put me on the horns of a dilemma, because it was my duty to prevent the use of tobacco in the abattoir, but clearly this hard-nosed Yorkshireman was superior to me by virtue of his qualifications. In the event, I chose to ignore the clouds of sweet-smelling cigar smoke knowing that once he departed I might be confronted by several slaughtermen in a state of justifiable high dudgeon as to why he could smoke and they couldn't. Fortunately for me, as I remember it, no such confrontation took place.

Having waited while I inspected all the lambs and all the offals, which might number as many as 200 at a time, this vet would then wait again while I stamped the carcasses with my numbered Bristol City Council health stamp. He would then produce his veterinary export stamp, and borrowing my ink pad, and without having actually checked a single offal or carcass himself, proceed to stamp every carcass as fit for export.

This early experience led me to believe that this particular level of bureaucracy was totally unnecessary, and extremely costly to the meat industry, an opinion which has not changed one iota in the intervening 45 years. Be that as it may, this particular vet subsequently became a good friend and we met professionally on many occasions in the years to come.

CHAPTER 5

ANTHRAX!

THE SENIOR MEAT inspector for the whole of Bristol, whose position I eventually achieved, Les Mawditt (who only died in October 2010 at the ripe old age of 99), took me under his wing, and as a result of accompanying him around the city I gained invaluable experience in the many and varied aspects of the job.

We visited the Bristol and Avonmouth docks to inspect imported meat and offals, butchers' shops and supermarkets, cold stores and meat depots, knacker and bone yards, private houses to follow up complaints, pet shops and piggeries. We also took samples for analysis from many of these premises, together with sewer swabs from the three abattoirs and the knacker yards, the results of which were detailed in an annual report for the perusal of the council.

We found organisms from all over the world, particularly salmonellas which would be identified by the city analyst, a process known as "phage typing". These had names usually related to where they were first found, such as Montevideo found in imported horse flesh for pet food from South America, Adelaide in kangaroo meat also for pet food, and also salmonella typhemurium which is hosted by rodents such as rats and mice and in one case grey squirrels (the latter causing a food poisoning outbreak in a residential home where they were being fed on the kitchen work surfaces by the staff). We even sampled the fundamental orifices of imported tortoises which were crammed into wooden orange boxes with total disregard for their welfare. Many of these were found to be full of salmonella and were destined to become children's pets.

The inspections at the docks were often associated with fridge breakdowns on the voyages. I remember one boat which had a cargo of frozen New Zealand lamb; it had had a fridge malfunction which had been corrected at sea. The bottom layers of lambs had thawed out and the huge tonnage of lambs above them had squashed them completely flat; many had to be rejected from this cargo. On another occasion frozen lambs had been stored in the adjoining hold to chilled citrus fruit and the lamb fat had absorbed an orange flavour, but with the use of amazing machines called ozonators this taint was able to be removed and the cargo saved.

By a strange twist of fate, many years later, my own son worked as an Environmental Health Officer both at Avonmouth and later at the Royal Portbury docks, to an extent following in his father's footsteps.

Les and I would be called to butcher's shops and supermarkets, mainly to sort out problems after fridge breakdowns, but occasionally to re-examine a carcass or offal which the butchers had found suspicious.

Cold stores would on occasion inevitably have problems with their refrigeration, resulting in straightforward decomposition of the meat, which would then be totally rejected. Where the temperatures had fluctuated in the storage period, various coloured mould growths would result. I remember one occasion, when I was senior meat inspector, being called to a cold store at Avonmouth where a huge chamber full of Irish intervention meat had been stored for four years. (Intervention meat is meat bought by the Government to remove it from the market in times of plenty and theoretically to return it to the market in leaner times.)

The meat comprised topsides, silversides and rump steaks, all beautifully wrapped and boxed – and covered in a black mould. My advice was total rejection of the whole chamber, but the decision had been taken, I think by the insurance company, to try to salvage what they could from the disaster by trimming. This was achieved by two men each opening each box, unwrapping each joint and sawing off with the aid of a bandsaw all the external surfaces. The meat was then rewrapped and put in new boxes. Inevitably, the men doing the trimming would touch the mouldy surfaces and then touch the new surfaces, transferring the mould spores from one to the other.

This process took many weeks and during that time I went in every day and randomly sampled 20 boxes to check the efficiency of the trimming. My advice was that each batch of freshly trimmed meat, once I had given the all clear, should be immediately used, but because it was intervention meat and technically belonged to the taxpayer, this did not happen. When the process was completed after many weeks the first boxes to be trimmed were rechecked and found to be again covered in black mould, whereupon I rejected the whole lot. What a waste of time and money!

In 1968 horrendous floods hit Bristol and the surrounding areas. Tremendous damage was done in certain areas of the city and there was much condemnation of foodstuffs as a result of direct contamination by flood water, or damage to refrigeration equipment by the water. The Bristol docks overflowed into the subterranean chambers of a building housing a cold store which contained many thousands of pounds worth of cheddar cheese, bags and bags of hops, bags of dried powdered egg, much of which was floating in the disgusting water.

It took several weeks for the clear-up to be completed, due to the complexities of the various insurance companies involved, and during this time a major mouse invasion took place, which spread to the frozen chambers higher up in the building. There must have been thousands of mice in there, mostly living behind the wooden walls in the cork insulation. However, some it seemed were able to withstand the freezing temperatures and were actually living and breeding in amongst frozen chickens. They would chew up some of the cellophane wrappers and make a nest of it inside the body cavity of the chickens. If I had not seen it with my own eyes I would not have believed it.

To eliminate this problem it was necessary to completely close the premises, which was then sealed and carbon dioxide gas pumped throughout the building. Afterwards all the contents had to be examined and either condemned or salvaged.

The meat depots in Old Market Street in Bristol sold mostly imported chilled and frozen meats, but some home-produced meat as well. Their business was carried out in the very early hours of the morning and there seemed to be an assumption that council officials would still be lying in their beds at this time. I knew the majority of the managers and salesmen at these depots personally, due to visits normally much later in the day to reject various meat and offal they had found to be defective. I also knew that many of these folk were smokers, and that when the cat's away, the mice will play.

Every once in a while I would make a very early start and secrete myself in a doorway, on the very wide street, and wait until somebody lit up a cigarette whilst cutting up meat for a customer. I would then stroll across the street in the darkness, but would inevitably be spotted. By the time I reached the premises the offender would have hidden the cigarette inside his hand inside his pocket. I would stand close to him and engage him in conversation, knowing that all the while the cigarette would be burning closer and closer to his fingers. Eventually he would be forced to withdraw his hand and the offending cigarette, whilst muttering some expletives about "bloody inspectors"; but the point was well made, and they never knew when I would be watching again.

There were two knacker yards operating within the city at that time. These would collect dead and dying farm animals from the surrounding area, a practice which I always thought was a great risk to public health, bearing in mind that some of these creatures could be injured in some way, but mostly because they had been suffering from some sort of terminal disease. Of the two knacker yards in Bristol, one of them supplied meat to pet shops which only necessitated the cooking of the meat and offal prior to sale. The cooking was done in big cauldrons in an adjoining room to where the dead animals were dressed and cut up, but the risk of bacterial cross-contamination between the cooked and raw meat by the operatives was extremely high. The housewife who ultimately purchased some of this product would be oblivious to the potential risks of contamination of human food by the use of cutting boards and knives both for this suspect product and later for food for human consumption.

The other knacker yard only used the carcasses of fallen stock for rendering, a process primarily used to produce tallow (rendered animal fat), which is employed in a multitude of industrial processes. Once the tallow has been extracted the remaining material was known as "graves" and was ground up and either sold as plant fertilizer (meat and bonemeal) or used as a constituent of farm animal foodstuffs; a practice which had been going for generations but became very contentious in later years – but more of this later.

On one occasion very early on in my career I was to see anthrax, something very few inspectors ever see. We were taught as students that the first symptom of anthrax in a ruminant animal was sudden death. On the first of these occasions I was visiting a yard which supplied pet food, to collect and renew the sewer swab. As I was driving down the lane to the yard who should be coming out but the council vet? He rolled down his car window and said "Hello Dave. Looks like we have got an anthrax, you should see this!" Whereupon he returned with me to the yard and we decided that I would leave my sewer swab in situ, because we had potentially the most dangerous possible scenario. The bovine animal had been opened and partially dressed before the knacker man concerned had decided that the organs, particularly the spleen and the blood, were very abnormal, and he had called in the vet.

The greatest danger with anthrax is not so much the bacillus in the carcass but the spores formed by the bacilli which, as the blood and body fluids dry, float off into the air. One can become infected by ingestion, inhalation or by inoculation via a cut. The first two of these routes usually prove fatal. The latter takes longer to get into the system, and can be treated with a fairly simple antibiotic. The bacteria can be killed fairly easily by high temperature, but the spores are practically indestructible, and only a naked flame will do the job. There is an island off Scotland called Gruinard where in 1942 the Government of the day experimented with germ warfare; for 50 odd years it was a no-go area until an English company was paid half a million pounds to decontaminate it. In 1986 the topsoil was removed and 280 tonnes of formaldehyde in 2,000 tonnes of sea water was soaked into the 520 acre island. Formaldehyde is unloved by the anthrax spore. There are, however, schools of thought who do not believe that the island is now safe bearing in mind anthrax spores have been found alive in a medieval archaeological site in Edinburgh.

Because of the great danger to the man who had opened the animal it was thought necessary to vaccinate him by antibiotic injection straight into his liver. Fortunately he did not become infected. However, everything combustible in the knacker yard – which included aprons, wooden-handled knives, cleavers and saws, all the other meat and carcasses on the premises at that time and obviously the anthrax carcass itself – were placed on plastic sheets in a concrete yard outside and ignited with several flame jets from butane bottles, all in an effort to kill any spores which may have formed. (I was given the job of remaining at the yard to oversee this process which took many hours). Eventually everything combustible had been reduced to a pile of ashes. The building itself was subjected to a thorough going over with a naked flame, which had to be done with great care to avoid setting the whole place on fire. Many years later I was reminded of this incident while overseeing some fires during the foot and mouth epidemic of 2001, about which I wrote in my previous book.

The second knacker yard in the city was a small part of a much larger bone yard. One day Les Mawditt and I were called to a suspicious-death cow carcass, which had been collected from a farm where it had inexplicably died very suddenly. Fortunately, this time the carcass had not been opened but was beginning to swell in the early stages of decomposition, causing a certain amount of blood and body fluid to exude under pressure from its various orifices. The instruction when anthrax is suspected is to plug these orifices with cotton wool, which we did, having taken a blood sample for testing at the city analyst's laboratory. This, at that time was in a building called Canynge Hall situated near Clifton Down station in Bristol, a place I visited frequently, when taking various samples for analysis including sewer swabs.

On this occasion, having left strict instruction at the bone yard that the carcass must not be touched in any way, Les and I both went to deliver the blood sample to the city analyst. With the benefit of hindsight one of us should have remained with the carcass because the sample turned out to be positive.

Returning in haste to the bone yard to organise the destruction of the carcass we were horrified to discover it was no longer in evidence and had been cut up, minced and mixed with approximated 20 tonnes of other material now residing in a huge cooking vat. Clearly, our instructions had not been circulated to all members of staff.

The rendering process had already started and would take about 24 hours. Once cooked the material was mixed with benzene, which facilitated the removal of the tallow. The benzene would then be tapped off and reclaimed for the next batch. There was nothing we could do but wait for the cycle to be completed. When at last the benzene was back in its vat, the tallow tapped off into metal drums and the graves tipped in a huge steaming pile on the floor, we started to sample every component part. Luckily for us no trace was found of the anthrax bacillus. This experience (however embarrassing) taught me that this was a very good sterilising process, when conducted for a long period at very high temperatures, not just for anthrax but for many other organisms.

CHAPTER 6

THE PIGGERIES

OCCASIONALLY A COMPLAINT would be made by a member of the public to the Public Health Department with regard to a meat product. I would be given the job of visiting the complainant to assess the problem. The problems could be as a result of bad handling somewhere in the chain from the abattoir to the consumer, or from poor inspection practices, and sometimes from a blatant disregard for good inspection practice. On these occasions, blame could be apportioned and prosecutions sometimes resulted. On other occasions, however, the problem could not be attributed to any particular individual; for example, an injection abscess in a piece of rump steak. This would be caused by bacteria from the skin being carried deep into the muscle by an injection needle, used either by a farmer or a vet. The resulting abscess would not necessarily be seen by either the meat inspector or the butcher, and might not be evident until it arrived on some unfortunate's plate.

Farmers would routinely inject piglets with iron, deep into the back leg to combat piglet anaemia. Abscesses could often form as a result. The leg could then be boned out, cured and then cooked to make ham, and I have seen pounds of ham which have been sliced on a machine where there was lovely pink ham around the outside and a spherical green area in the middle which in fact was cooked pus. However undesirable and unpleasant this may appear, it was probably totally sterile due to the cooking process, which involves a high temperature for a long period. In these cases I would have to try and placate the consumer as there was no chance of apportioning blame, but would make it my business to visit the butcher or supermarket manager concerned to suggest a consolatory offer of a generous replacement to the customer.

Wednesdays were usually a quiet day at the abattoir and I often used the afternoons to visit one or two of the many piggeries which were within the city limits. Some of them operated on council-owned allotments, whilst others were on privately owned ground. Many were hobby farmers, keeping a few pigs for themselves, while others had developed into fairly big commercial operations. My role was to check the movement licences, which were issued every time a pig was moved from one holding to another. A copy would be sent to the local council and I had to check that the movement had been completed, the numbers were correct, and that there was no evidence of infectious disease.

Most, but not all, of these farmers were swill boilers. The smaller ones would collect food waste from perhaps one or two shops, whereas the larger ones would have rounds, collecting from big commercial canteens, school kitchens and the like. All this food waste was taken back to their premises and cooked. The larger operators had huge Barford boilers which produced jets of steam to be injected into large tanks which contained the swill. The smaller people often used an old-fashioned laundry copper heated by gas from a cylinder. In most cases the process was carried out very efficiently, but as with any other walk of life there would be those who would cut corners and would feed the swill without any sterilisation. The potential danger of this malpractice lay with any meat and bones in the swill, which could possibly transmit infectious diseases such as swine fever or foot and mouth to the consuming pigs.

At that time, large quantities of meat on the bone were imported from places such as Brazil and Argentina, countries in which foot and mouth was (and still is) endemic. In fact, some outbreaks of disease in the UK were considered directly attributable to swill boiling, and later on meat from these countries could only be imported in a boneless state, the theory being that the infection could live deep within the bone and resist the boiling process. Eventually the feeding of waste food and swill boiling was banned altogether. At that time, however, my job was to see that swill boiling was carried out correctly.

As I write, many local authorities are issuing food waste bins to households with the intention of composting the contents. This is obviously very commendable, to lessen the bulk going to landfill, but surely it would be better to utilise all these thousands of tonnes of food waste which could be properly cooked and sterilised at a central point in each authority to produce a pudding which could be cheaply sold to pig and poultry producers? This suggestion is inevitably going to be controversial to many people, and I would agree with the concerns if it were done on an ad hoc basis; but properly done, it could be very useful. Indeed, at the time about which I am writing some local authorities made pig-bin collections from every household, the collections of which were treated as I have described.

On a lighter note, many of these piggery owners were really interesting characters. One, unbeknown to the council, had built a second storey above some of his pig houses, in which he installed all the necessary facilities and spent much of his time living above his pigs. I had been visiting him for some time before I was deemed trustworthy enough to be invited up to his inner sanctum for a cup of tea and a chat.

Several of the others, not wishing to spend money on fuel for their swill boilers, sourced an unusual form of fuel from a boot and shoe manufacturer in Bristol called GB Britton, who made the well known "Tuf Boots". They acquired vast quantities of reject footwear and burned them on their solid fuel boilers to make the steam to cook the swill. Inevitably, this was accompanied by clouds of acrid black smoke in the vicinity, but as there were no residential properties close by, nobody seemed to mind.

This of course was in the days before the Clean Air Act, since then things have rightly been tightened up. I well remember during the 2001 foot and mouth outbreak, with which I was deeply professionally involved, that the funeral pyres had to come to an end because of concerns about dioxins beings released into the atmosphere. Goodness knows what was coming off these burning boots and shoes!

All of these piggeries were by their nature very susceptible to vermin problems. Most of the farmers took some steps to limit the populations, but one particular man seemed oblivious to the fact he was overrun by rats. At any time during the day, if you visited, you could see dozens of rats blatantly running around right under your feet. The problem was so bad that Les Mawditt, with whom I was visiting the piggery, insisted on calling in one of the council rat catchers (now politically correctly called rodent operatives) to poison-bait the premises. Vast quantities of poison had to be used, and when we visited again some days later there were rat carcasses everywhere. This man was equally unconcerned about the dead rats and made no effort to remove them. When Les suggested that he should pick up the carcasses and perhaps burn them in his boiler this individual took offence and picked up an axe he used to split wood, held it aloft and threatened us, telling us to get out. I surreptitiously reached for a piece of timber with which to subdue this man should it become necessary, but in the event Les stood his ground and the axe was lowered, and finally dropped. One wondered after we left how many rat bodies were consumed by his pigs, complete with the poison inside them. I heard some years later that this fellow had died of Weil's disease, a particularly unpleasant infection which, as you may know, is to be found in rat urine and faeces, and carried by about 80% of rats – poetic justice comes to mind.

CHAPTER 7

WASTE NOT, WANT NOT

THE SECOND OF the three meat plants that operated within the city limits at that time was, as I have already mentioned, a factory specialising in pork and bacon products. Les Mawditt and I attended there on Tuesday, Wednesday and Thursday mornings, to inspect their weekly kill of about 1,000 pigs.

The pigs to be killed on each day arrived the night before and were kept in the lairage overnight. This served two purposes: first, it allowed the pigs to relax after what was probably a stressful journey, and secondly it allowed them to largely empty their digestive system. The downside was that the consignment from each farm may have included pigs from different pens on the farm, according to their fitness for slaughter, and even though these pens may have been adjacent to one another, when confined together territorial fighting would often occur. Mostly the tooth damage was fairly superficial, but would spoil the visual aspect of the carcass, but on occasion (when the beaten animal could not escape from the confines of the pens) death could occur from exhaustion rather than injury.

At the start of slaughter, as each batch of pigs was brought into the stunning area, they would be hosed off, partly to remove the gross contamination of excrement which may have occurred overnight and partly to improve the electrical contact when the stunning took place. In fact the electrodes of the stunning equipment would be kept when not in use in a bucket of saline solution for the same reason. This procedure was very commendable on both counts, but there was a big disadvantage. When the pigs were "stuck" an immaculately clean stainless steel bucket would be held under the throat of each pig to collect the blood, for the making of black pudding. Unfortunately, sometimes, when the pig kicked in reflex action, or if the concentration of the slaughterman holding the bucket lapsed for a moment, it was possible for the drips of the washing water – now probably containing nasal discharge, faeces, urine and semen – to go into the bucket together with the blood. The addition of semen to the other body fluids was as a result of many male pigs ejaculating after electrocution.

If these unwelcome additions to the blood were noticed the whole bucket would obviously be rejected, but I am sure on many occasions they added to the flavour of the black pudding! Needless to say I have never eaten black pudding since. It has to be said that there is now a hollow bladed knife system which sucks the blood into a sealed container; nevertheless, these early memories of mine have stayed with me to the present day.

The buckets of blood contained an anti-coagulant, and each churn represented so many pigs, so that if later down the line of the dressing process an animal was found to be totally unfit for human consumption and was rejected the appropriate churn was also thrown away.

After death the pigs would be immersed in a scalding tank to loosen the hair follicles and two or three layers of skin. They were then lifted into a dehairing machine, and rotated by rubber paddles with blunt steel blades attached. This removed the loosened skin and hair. The pig would then be tipped onto a table where the tendons at the back of the trotter would be exposed with a knife and a gambrel would be inserted behind them. The pigs would then be hoisted onto a bar on which they travelled into an oil-fired burner composed of a huge vertical steel cylinder in two halves which opened and closed on rollers, operated manually with a lever. The two halves were clad inside with fire bricks. The purpose of the burning process was to soften and seal the skin so that it could be sliced on a bacon slicer. The final process before evisceration was to remove the blackened surface of the skin, leaving it a golden brown colour. However, some of the smaller pigs would not be burnt in the manner described, but would be flashed very quickly through the burner to remove any remaining hair, then remaining the normal white colour of pork pigs. The bodies would then be eviscerated. The large and small intestine and the stomach were carefully removed and placed in sequence on a table where they would be inspected. This involved the incision of *every* lymphatic gland appertaining to the stomach and the intestines primarily looking for signs of tuberculosis, be it bovine or avian type. The "pluck" (liver, heart and lungs) would be removed and similarly inspected; Les and I used to alternate, one day inspecting carcasses, the next red and green offal. These jobs were done within easy shouting distance of each other so that should something untoward be found in either the carcasses or the offals we could inform one another of our various findings and take appropriate action.

Much of the early preparatory work for processing of the pigs into bacon or pork products was done in the abattoir. This consisted of head removal, splitting of the carcass and the stripping of the peritoneal (abdominal) fat. The pigs would then be left in a hanging area to set and cool until the next day, whereupon they would be dispatched to various areas of the wider factory depending on whether they were destined for pork or pork products. Bacon pigs would be partially boned out and the pork steaks (tenderloins) and kidneys removed to be sold fresh. In addition, some so-called "heavy hogs" would be slaughtered, together with a few adult sows and boars. The "heavy hogs" were usually swill-fed and little import was given to their body conformation, as opposed to the pork and bacon pigs which had to be just right on the fat-to-lean ratio. These heavy pigs were destined to have the majority of their fat and skin removed.

A speciality of this firm at the time was something called a "Griskin loin", which stretched from behind the head right down to the back. The spine bone was removed as was the shoulder blade, but the rib bones remained. Just the right amount of fat was left on the outside and the whole length could be cut into very succulent chops. The remainder of the pig consisted of the leg, which would be cured for ham, and the belly and lower half of the shoulder, which would be used for manufacturing, together with all the meat from the adult sows and boars, the latter considered to be too tough for any other purpose.

In the wider factory, where probably 150 people worked, several further processes were undertaken. The first was the curing of bacon sides. This was done by injecting brine cure into the deep muscle of the legs and shoulders. The sides were then completely immersed for a specified period in large tanks like small swimming pools to cure throughout. The tanks were situated in the cellars of the premises and had obviously been curing pigs for so long that stalactites had formed on the ceiling above. Having been cured, some sides remained as "green" bacon but some would be put through a further process of smoking. The sides were suspended on a bar in large walk-in smokers. Oak chips were filtered into a boiler and the resultant smoke was piped into these chambers for a designated number of days. When the ovens were opened a very pleasant wood-smoke smell would pervade the whole factory. In fact my wife could sometimes smell it on my clothes when I arrived home.

In another process the waste fat taken from the heavy hogs, together with all the belly fat from the bacon pigs, would be minced and put in a steam-jacketed cooker with paddles rotating inside to render down the fat from the other tissues. This liquid fat, by then known as lard, was poured into greaseproof bags, some to take half pounds and some to take one pounds, which were supported in wooden racks until the lard had set. Some of these came my way when the factory closed and are still used as tool racks.

"Bath chaps" were also made. These consisted of one cured cheek and half of the tongue, cooked in a conical mould, tipped out when cold and covered in golden breadcrumbs. These went out of favour when people were advised to limit their fat intake, but recently some of the celebrity chefs have been highlighting the succulence of the cheaper cuts of meat, including pig and cattle cheek.

In another area of the factory, Spears pork pies, famous throughout the South West, were produced. These were made in various sizes ranging from a small individual to a large round family size, and also the long veal, ham and egg pies, which contained to my knowledge no veal at all! Later on, when the Trade Descriptions Act came in, they were re-named "Festival pies".

Sausages and black pudding were also made, together with a meat sausage in a red plastic skin, called a "Polony". The sausages were second to none.

During all these factory processes there was a certain amount of waste, ie damaged pies, sausage meat left over from a batch or trimmings off the hams, which would all be minced again and put into the Polonies. This of course was in the days when everything was used, not like today!

The pig red offals (heart, lungs and liver, aka 'the pluck') were made into the most delightful faggots which were wrapped in the omentum (caul or abdominal) fat. The green offals, in particular the stomachs and small intestine, were thoroughly cleaned by turning them inside out to remove every last vestige of faeces, and also to scrape off the mucus lining of the gut. This was done over the side of large wooden tanks filled with hot water. The small intestine was then plaited, salted and cooked, to become chitterlings. The stomachs, known as maul bags, would also be cooked, to be sold together with the chitterlings, both of which were highly prized as a delicacy, particularly in pubs to accompany beer. However, the real connoisseurs of chitterlings particularly sought out what was known as the "fat end", which was the rectum and anus, cleansed salted and cooked in the same way. The remainder of the large intestine – the caecum and the colon – could be cleaned, but this was very time-consuming and not often done.

The connective tissue which was left after the various fats had been rendered for lard was sold as pork scratchings. The trotters and tails would be salted and cooked. Pork bones (the removed spine from the bacon pigs) had a ready market, and even the ears were sold for human consumption. More recently, dried pig's ears are sold for animal chews, and River Cottage chef Hugh Fearnley- Whittingstall revived a recipe on one of his culinary programmes using pig's ears. The old adage that you can use everything from a pig except the squeal is absolutely true.

All these products were sold to local butchers, and in particular speciality pork butchers, one of whom still remains in Bristol. Over time this shop has become surrounded by a population, many of whom are of the Muslim faith, whose sensibilities are apparently offended by the sight of pork products in their midst.

This shop restricts the advertising of their traditional delicacies just in case they might offend people many of whom have only comparatively recently come into our community and in my opinion have not the right to influence the established likes and traditions of the indigenous people. We all need to live together in this world and this means that we need to curb our sensitivities so people can carry on with their traditions – and for many of us that is the eating and necessary displaying of pork products. No one's tender sensibilities should be so easily offended to interfere with the likes and traditions built up over so many generations. Just possibly people like me may be offended by comparative newcomers having the audacity to challenge my inherited likes and traditions which are being eroded away on a daily basis.

I also read in the paper recently that this shop has discontinued the production of "bath chaps". The proprietor had decided that the struggle against bureaucratic madness which had descended upon their production was just not worth bothering with.

Once we had finished inspecting the daily kill we would adjourn to a small office which had been put at our disposal, having first visited the works canteen kitchen to collect a cup of tea, and also having called in at the pie room to sample a hot pie straight out of the oven before the gelatine had been injected into them. I can still taste those pies now – I have never tasted better.

The Spears factory was situated opposite a building site and from our little office we observed the progress of the construction of the new building to house the *Bristol Evening Post* and the *Western Daily Press*. Sadly, in the 1970s, with no apparent heir to continue the family business, the Spears factory closed. The name was bought by some of the managers and they continued producing some small goods in a modern factory unit, but in the end even this eventually disappeared.

CHAPTER 8

GORDON ROAD

THE THIRD ABATTOIR, where I spent most of my Bristol working life, was situated at Gordon Road, in Whitehall. The abattoir was owned by Bristol City Council, as were many at that time, when there was an obligation for councils to provide slaughtering facilities for the local butchers.

The day-to-day running of the plant was often delegated to a slaughtering contractor. In the case of Gordon Road, this contractor was a lovely man called Roger Hendy, who became a very good friend of mine. When the obligation on councils ceased, many abattoirs were sold to the private sector and in this case Roger purchased the premises in partnership with the largest wholesale butcher using the plant, Ken Hill.

When the council still owned the plant Les was both senior meat inspector and abattoir manager, assisted by a clerical officer. Next in command was Teddy Howick, another lovely man who was like a father both to me, as a youngster, and to the many public health, meat inspector and vet students who trained at this abattoir over the years. Both Teddy and Les were public health inspectors. Les was originally a plumber, an occupation from which many public health inspectors were sourced. Teddy Howick's background was a bit obscure, but I do know he worked in the Port Health Department at Avonmouth before his time at the abattoir, where he stayed until his retirement. Teddy lived in an old manor house on the outskirts of Bristol where he would hold fantastic parties. Among the guests would be interesting people like clarinettist Acker Bilk, Adge Cutler and The Wurzels.

Intermittently other public health inspectors, who would normally be doing district work in their various capacities (eg housing, food or pollution control) would stand in at the abattoir doing meat inspection to cover sickness and the holidays of the regular meat inspectors. In those days all these men and women were very highly trained in all aspects of public health work. Sadly today meat inspection, in what has become part of the environmental health course, has taken very much a back seat, and lip service and a rubber stamp is all that is deemed necessary for what is in my opinion a very important aspect of public health.

There were also two meat inspectors, and an assistant. Working with me was Les Worden, with whom I trained. He had previously been a manager at Hotwells. There was also Louis Thorn who had been a butchery manager for the Co-Op before becoming an inspector, and who had previously served in the British Army in Palestine. Louis used to regale us with stories of the horrendous torture and mutilation of captured British troops by the terrorist factions of the day or so-called "freedom fighters". It is interesting how one day's terrorist becomes next day's politician, both back then in Palestine and more recently in Northern Ireland, where people with questionable legacies hold high office. Many other examples exist throughout the world.

There were two cattle gangs and one sheep gang, each consisting of five men all in one building; in a separate building just two men coped with the entire pig kill. The inspectors would alternate between the various lines as carcasses and offals became available for inspection. Sometimes this would leave me with time on my hands, which I put to good use helping various gangs by slaughtering, initially with a provisional licence under supervision, and later becoming a licensed slaughterman in my own right. I have always believed that when in a position of authority one should not ask someone else to do something one is not prepared to do oneself.

Part of my job was to assess and criticise the expertise of the slaughtermen, and to ensure the welfare of the animals at the point of slaughter was as good as it could possibly be; I gained the reputation that I would not tolerate less. In fact, the only prosecutions I have ever brought in my life were against people who were persistently cruel to animals.

Both Les Mawditt and Teddy Howick were very good at helping and schooling new inspectors. I cannot say "young new meat inspectors" because, although I was about 20 when I was trained, Louis Thorn and Les Worden were considerably older. Teddy Howick in particular would take us carefully through any unusual disease or condition which had come to light. As a result of this careful tuition our knowledge and expertise increased by the day. It has to be said that whereas we started work at seven am, Teddy would wander in at about half past eight, a copy of the *Sporting Life* tucked carefully under his arm and a Senior Service gripped between the first two fingers of his right hand. He would come to the door of the cooling hall, which was a building separate from the main slaughter hall. Interconnecting rails with roller hooks joined the two buildings, and sides of beef would be washed with a high pressure hot hose and pushed across onto the scales in the cooling hall. Here they would be weighed, reunited with their other half and put in rows on the many rails within the building, allowing a slight gap between them to let the body heat escape. Sheep and lambs were hung four at a time on what were known as "stars". These were long steel rods with a ring at one end to go over the hook on the rail and four hooks at the other end forming a star shape. The pigs were hung on individual long big link chains up to weighing and were then transferred to stars.

Teddy would call out to us to ascertain whether there had been any problems up to that point. Having satisfied himself that everything was running smoothly he would then return to his car to collect some eggs, bread and butter. He would then retire to the mess room where the bread was toasted and buttered, and the eggs poached and placed upon it; it never ceased to amaze me how he did not become egg-bound. He then got to down the serious business of the day, which was to see which were going to be the lucky horses to have his money placed upon them. He would then return to his office and ring up his bookie. Occasionally, he would have received a tip, or particularly fancy a given horse selected from the *Life*, and in these cases he would generously impart the necessary information to the rest of us so we could enjoy a flutter.

At around 10 o'clock, when the rest of us were just going to breakfast, Teddy would have donned his protective clothing ready for the day's endeavours. The first thing he would do would be to walk between the rows of carcasses, stopping once or twice having seen some minuscule problem which evidently one of us had missed. We would be hauled in front of him like naughty school boys to be reprimanded, and reminded that we were there to protect the public health. His favourite expression when pointing out one of these errors was "What do you think this is, Scotch mist?", a phrase I have used myself on many occasions in the intervening years whilst teaching students who once upon a time were allowed to practise meat inspection under the supervision of a qualified officer. Sadly, this is no longer the case.

Teddy would then do a bit of meat inspection, relieving each one of us in turn so we could have a tea break. The system only stopped for a breakfast break and a lunch break: the rest of the time we carried on killing.

In any other circumstance, Teddy's shorter working hours could have caused resentment among the rest of us, but I never heard a word of criticism in all the years he worked with us. In fact, he was something of an institution after so many years in the job and all of us were very fond of him.

My life now settled down to a routine, all the while gaining invaluable experience. On the three middle days of the week I continued with Les Mawditt to cover the bacon factory, which was a slightly later start at half seven. When we had finished the kill at Spears, Les would take me with him around Bristol, following up various meat-orientated food complaints and other assignments. I remember one day we had a complaint about illicit slaughter being carried out in a premises in St Paul's in Bristol, where took place the riots in 1980. We initiated a raid of the premises where police forced an entry for us to look for any evidence of the illicit practice. There were in fact no carcasses present on the site, nor any live animals, but there were quite a lot of skins, also blood and hair which would seem to confirm our suspicions. Unfortunately, the perpetrator jumped over the back wall of the garden and disappeared into a maze of alleyways and gardens.

A few days later an Afro Caribbean gentleman appeared at the abattoir demanding to see Les Mawditt who was doing some paperwork in his office at the time. This guy burst into the office and started to verbally abuse Les very loudly. We in the slaughter hall quickly became aware of what was happening and ran to Les's assistance. Confronted by three bloody figures heavily armed with knives in pouches, the man decided discretion was the better part of valour and beat a hasty retreat.

Sadly, illicit slaughter is probably more common now than then, and today's officialdom is completely out of touch with what is going on because they do not have the network of local knowledge that Les and Teddy possessed. The increase in numbers can be attributed to the increase in several ethnic populations, particularly the sub-Saharan Africans, who are especially fond of what are known as "smokies". These are animals which have their hair burnt off them before evisceration, which probably gives the meat a particular flavour akin it could be said to smoked bacon etc. I have no problem with this but the law of this country says you will flay (skin) a sheep or goat carcass and if it is going to be offered for sale or maybe as a prize in a raffle it must have been killed in a licensed premises by a licensed slaughterman, properly inspected and stamped with a health mark.

Many times over the years it has come to my attention that goats, particularly young goats, are obtained privately by people from both the Indian sub-continent and the Caribbean and killed illegally in back yards and gardens. This also applies to poultry obtained in poultry sales or directly from the producer. About 10 years ago I used to attend all the local bird sales and there was one Asian gentlemen who used to buy all the cockerels that people would bring to the sales hoping to find a good home for them. The owners were unaware of the fate that awaited these unfortunate birds, perhaps ending up in a curry sauce.

Clearly all these illegal practices are extremely difficult to police, but as well as policing the issue it needs to be raised and openly and freely discussed. It seems to me very few are prepared to raise a voice in protest about anything to do with ethnic minorities for fear of being accused of racism or religious intolerance, regardless of what the law of the country says.

CHAPTER 9

LES'S MISHAPS

WHILE THE COUNCIL still owned the abattoir, and Les was the abattoir superintendent, if we finished early on a given day he always seemed able to find me something around the premises to do. This would range from grass-cutting to the spraying of weeds, or filling bags of manure to improve his immaculate garden.

The grass cutting started out using an Allen scythe with scissor-like blades which was an extremely heavy and unwieldy tool, especially when used in a small area. Eventually Les invested in one of the new-fangled Flymos and I was thoroughly schooled in the operation of it, particularly on steep banks, of which there were several. This process necessitated tying a piece of rope to the handle, standing at the top of the bank and lowering the machine to the bottom, then hauling it back using the rope. Les's first demonstration nearly ended up badly. Pulling too hard on the rope he brought the mower over the top of his foot, removing some of his boot and his sock. Fortunately, none of his pinkies were damaged, and only his pride was hurt.

On another day, against my better judgement because there was a slight wind blowing, I had to spray all around the perimeter fence of the abattoir to kill the broad-leaved weeds. Even though I did my best to keep the nozzle close to the ground the adjoining allotments (several of which abutted the abattoir fence) soon experienced an inexplicable die-off amongst their young peas and beans. It did not take long, however, for the affected gardeners to realise what must have happened, and Les received a deputation of them complaining, necessitating compensation to be paid by the council.

Events like this were the downside, but the upside of Les's mentoring was that I received a very broad experience, much more so then most new inspectors of the day. This stood me in very good stead throughout my career, especially when I became the senior inspector in Les's place.

Les took his job as the abattoir superintendent very seriously. He was very capable when it came to matters of mechanical maintenance, but on occasion things would go sadly wrong (bear in mind that this was in the days before health and safety legislation). By the side of the main evisceration table was a piece of machinery known as the "fat blower". The green offal room was situated on the floor above the slaughter hall, which necessitated all the green offals and stomach fats to be transported upwards. The large items, such as the stomach and intestines from cattle, were placed in a large tray after inspection, which was then hoisted to the required height. The smaller items, such as the sheep intestines, sheep stomach fat and cattle stomach fat, were placed in the blower. Copious hot water was added, and then two levers had to be pressed simultaneously whereupon a door would close the machine and compressed air would propel all the contents upstairs via an eight-inch plastic tube. This system worked very well as long as those operating it remembered to use plenty of hot water. Sometimes this was forgotten, however, and, especially in cold weather, the fat would start to set in the machine. This could result in a blowback into the slaughterhouse on the release of the levers, but it could also cause some fat to set half-way up the tube causing a blockage. At times like this Les would don his overalls and, with the aid of a spanner, remove a section of pipe to clear the blockage, having left instruction that whilst he was so engaged no-one must operate the machine.

On this particular day Les was working on a section of pipe out of sight of those below when one of the slaughtermen returned to the slaughter hall from elsewhere. Knowing there was a blockage, but not that Les was working further up the pipe, he said "I'll soon sort this out" and promptly poured two or three buckets of scalding water into the machine and operated the levers. There was a cry of anguish from behind the next wall, and Les appeared soaking wet and covered from head to foot with little pieces of sheep fat. His language was unrepeatable, and on seeing him in this condition a stunned silence descended on the slaughter hall, which lasted until Les had withdrawn from the building to try to clean himself up, whereupon the whole place erupted with hysterical laughter. Poor Les had no option but to return home for a bath. Fortunately, no lasting damage was done, and the only thing hurt was once again his pride.

There was also one occasion that I remember at the bacon factory when a big sow escaped from the lairage into the slaughterhouse. Les was bending over, deeply engrossed in examining a pig's head, and as he straightened up he became aware of this very large animal bearing down on him. He quickly raised his apron in an attempt to turn the sow back from whence it came. Unfortunately, the sow had other ideas and headed for the gap now visible between Les's legs, and by doing so lifted him clean off his feet and carried him for some distance astride her back but facing in the wrong direction. Even though Les was then unceremoniously deposited on the concrete floor he was unhurt apart from a few minor bruises, but once again his pride took a major battering as the slaughterhouse erupted into laughter. In my lowly position, however, I had to pretend not to notice: it's amazing how engrossing inspecting a pig offal can be!

CHAPTER 10

MORE GORDON ROAD

WHEN I WAS a very young inspector money was quite short, and as I was contemplating getting married to my long-time girlfriend Alison, any way of making a bit extra was very welcome.

Back at Gordon Road we were killing thousands of old ewes and rams for the "curry trade", and from December through to about May many of the ewes were heavily pregnant as there would have been rams running with them over the winter. Many of these ewes would have been purchased by the dealers during the previous autumn, when farmers would go through their flocks and cull all those which they considered to be at the end of their working lives. This could be purely through old age, or maybe some of their teeth had fallen out so they could not graze properly (broken-mouthed sheep). Possibly they were lame from bad foot-rot problems or arthritis or injury. Then there were those which had mastitis problems in their udders, sometimes of a horrendous nature.

The dealers would overwinter these sheep, drawing on them in the spring when there were very few ewes available because lambing was taking place on the farms, and the only animals coming forward for slaughter would be those that had lost their lambs, had never been in lamb in the first place or were suffering from some sort of problem as a result of lambing.

In the slaughterhouse the heavily pregnant uteri would be collected in 50 gallon (often rusty) metal drums, which in those days were the containers used for the collection of waste materials. It was nothing to see up to 10 of these bins a day full of lambs. I became aware from one of the drivers on the hide lorries that the little woolly skins from these advanced foetuses were saleable for the princely sum of 1 shilling each (5p in today's money). Having obtained permission from the owner of the sheep to remove these skins, I spent many lunch hours and break times so doing.

With a little bit of TLC some of these ewes could successfully complete another lambing season, so when I was not engaged in skinning lambs I would carefully go through the various batches of sheep in the lairage and separate out those which I considered were worth another go. Any animal with mastitis was a non-starter, but those with broken mouths and a few loose teeth left (but still in good bodily condition) would be able to graze on lush grass, having had these loose snags removed, enabling them to bite gum to gum in quite an efficient manner. Those with bad feet could sometimes be rescued with trimming and treatment.

This selection allowed me to build up a flock of 30 to 40 animals. I sought permission once again from the owner to remove them from the abattoir on the understanding that they would be returned when the lambs were ready for slaughter, and I would only have to pay for any ewes which might die during the lambing season. He would buy the lambs back from me, and the ewes would be returned usually in much better condition than when they left.

I was able to do all this because I had the use of 30 acres of land at Almondsbury which belonged to a man who was engaged to my sister at the time, and who had an early tomato farm under several acres of glass, and just wanted the rest of his land kept tidy.

The first year when it came to shearing time, I would transport six or eight ewes at a time in my little trailer about 10 miles up the M4 to where the butcher owner of the ewes kept sheep on his own farm. He would employ shearers to shear not only his sheep but also those of his I had borrowed. The problem was transportation of only six or eight at a time, and I would be driving back and forth like a yoyo.

Inevitably, something had to go wrong, and it did, in the form of a puncture in one of the wheels of the trailer. Having no spare wheel, and mobile phones not having been invented back then, all I could do was leave the trailer containing six ewes on the hard shoulder and drive as fast as I could to the shearing destination. I persuaded the farm manager to drive in a small cattle lorry back down the M4 (passing the sheep in the trailer on the other side), turning around at the next junction and then driving back up the M4 to where fortunately the six sheep and the punctured trailer were still waiting. Several of us reversed the trailer so that we could release the back door onto the gangplank of the lorry, a tricky and potentially dangerous operation. The transfer of the sheep was successfully achieved and having boxed them in the front of the lorry we pushed and pulled the stricken trailer up behind them.

This was an experience I shall never forget, and I shudder to think of the consequences had the sheep escaped onto the motorway, but it is amazing the risks you are prepared to take when you are young.

The next year a Welsh former hill sheep farmer came to Gordon Road to learn to be a meat inspector. Whilst walking around the lairage one day looking at the stock he was able to tell me from exactly which part of Wales any one of the Welsh ewes had originated. He was able to do this because, together with his father, and probably his grandfather before that, he could read the notches cut in the ears by the Welsh hill farmers to identify their sheep when rounding them up from the hills. Most of the Welsh hill sheep come from what are known as "hefted" flocks, which basically means that when they are turned out onto the hills they naturally remain all their lives within an area, usually comprising only a few square miles, in which they were born and raised. Occasionally, however, some sheep may keep walking and get lost, and at times of round-up it needed men who could recognise the ear coding at a glance and were then able to return an errant sheep to its home farm.

This man told me that when he had been hill farming he was able to shear by hand about 90 Welsh ewes a day. Although sheep farmers keep hand clippers to clip away contaminated wool from around the tails, the whole body is generally shorn with electric clippers. Ideally, the ewes would be first driven through a river which has the effect of making the wool rise away from the body, thereby facilitating shearing. I told him that I had about 35 ewes which needed shearing, and he said that as I had helped him with his practical meat inspection he would shear my ewes for me, as long as I supplied sufficient beer for him to drink while he did the job. True to his word, on the appointed day, which was a very hot Sunday morning in early May, he sheared my flock, using some exquisitely sharp hand clippers. He sweated profusely and drank copiously. In the end he did an incredibly good job under very difficult circumstances, because the wool had not risen and also because of the variety of breeds, from the little Welsh mountains through to huge Devon curly wools with wire-like wool. There is no way he would have done 90 of these in one day, but with me catching and him shearing we completed the job by lunchtime.

Prior to training as a meat inspector I worked in the butchery trade and was fully familiar with the intricacies of cutting up carcasses of food animals for the retail trade. When the shop in which I worked closed on Monday afternoons I would drive on my motorbike a short distance to the northern edge of Bristol, where there was a small abattoir which supplied not only a shop on the premises but also some shops in Bristol. My purpose was to assist the slaughterman and learn the killing side of the business. This stood me in good stead when, later working at Gordon Road, I was able, in between doing my own job, to help the slaughtermen with theirs.

Sadly today inspectors are forbidden to assist the abattoir owners in any way; to do so invites possible dismissal. The reasons for this are several: first health and safety concerns; secondly because it is feared that helping the abattoir staff could compromise the inspector's authority; and lastly because it seems today that officials view what are now known as food business operators as the enemy, and on no account must you assist the enemy. To my mind, this attitude is entirely wrong and engenders a feeling of "them and us". In my day we all helped each other, to put a good clean healthy product out to the consumer. We became a team where mutual respect was the order of the day, and physical help from the inspectorial staff was reciprocated by the abattoir staff occasionally pointing out defects which could potentially be missed. All this was conducive to a very happy working environment which seems to be largely absent today. Very few vets or inspectors today are licensed slaughtermen and most would not have the necessary ability or inclination to kill an animal, and even less be able to dress one for themselves.

As at the Hotwells abattoir, there were many real characters amongst the staff at Gordon Road, indeed a few of them transferred from Hotwells when it closed. Contrary to what might be some people's perception of slaughtermen, in my experience, with very few exceptions, they are decent, hard-working men with a tremendous, often black, sense of humour. I have many friends amongst the slaughtering fraternity and they have in large part contributed to my very pleasurable working life. I think if you can look back over your working life and can say that for the most part you have enjoyed it, you are very lucky, and I am a very lucky man.

I would like to recount a few anecdotes from those days at Gordon Road, some sad, some painful and some funny.

I will start with Alfie James, an older man who had been slaughtering all his life. He had sustained a serious head injury I think in the Second World War, and as a result had a steel plate in his skull. Unfortunately, this had left him suffering from epilepsy, and with very little warning he would collapse wherever he happened to be, and we would have to try and grab him when he was thrashing about in the blood on the floor to stop him from damaging himself on the many sharp objects and solid pieces of equipment. Sometimes his tongue would fall into the back of his throat and had to be retrieved by inserting a less than hygienic finger into his mouth. After one of these incidents poor Alfie seemed to find it necessary to apologise to everyone for the disruption caused, and more often than not, because he did not drive, he would have to be taken home to clean himself up.

In the very dangerous environment in which we worked – sharp knives, sharp hooks, sharp saws, large animals, overhead moving equipment, slippery floors, you name it – it was inevitable that accidents would sometimes happen. The worst one I can recall happened during a time of refurbishment when a temporary cattle dressing line had been constructed in the pig slaughter hall. A cattle-stunning trap had been erected with a long race leading to it made out of scaffold tube. Bill Pratt, who was the foreman of the beef gang, spent all day driving cattle into the stunning pen, shooting them and then going back to fetch another one. If an animal decided it did not want to go into the stunning trap it would walk backwards and force the man driving it back down the race. Sticks were in common use in those days, electric goads having only just been invented. Bill was a very gentle older man and preferred to use his voice in quiet tones and a persuasive slap on the rump with his hand, which seemed to work very well in most cases.

On this particular day we had a load of Friesian steers to kill. In those days most cattle had horns because disbudding was not generally practised. These steers had short straight horns about six inches long, set at right angles to their heads. One of these animals somehow managed to rear up on its hind legs, turned completely around in the narrow alleyway and returned at speed from where it had come. Bill had nowhere to go as he could not back-pedal fast enough to get out of the way. I saw from the slaughterhouse the animal returning at speed to the lairage and instinctively knew that if Bill was in its way he would be in trouble. I shouted to the others and several of us ran out into the lairage to find Bill on his back in the straw and faeces with a horrific injury. One of the steer's horns had pierced the side of Bill's neck just under his jaw, leaving his trachea (windpipe) exposed. Incredibly, it had missed his carotid arteries and jugular veins, otherwise he would have bled to death in a very short time. Unaware of how badly injured he was, and only semi-conscious, he implored us not to tell his missus what had happened so as not to worry her. Somebody found a clean shirt to wrap around his neck, and even though the four major blood vessels had not been severed, he was still bleeding profusely. In due course the ambulance arrived and Bill was lifted out of the manure onto a stretcher; just as they were closing the doors on him he asked "Has anyone seen my glasses?" I quickly returned to the scene of the accident and, poking around in the manure where everybody had walked, found the missing spectacles completely undamaged but covered in faeces. After a quick swill under the hot tap and a wipe on my handkerchief I quickly returned then to Bill's nose as he set off to hospital. When he returned to see us some weeks later he had seemingly made a good recovery, but had more or less lost his voice, which was surprising as he had clearly spoken to us in the ambulance. He never worked again.

During the time of this same refurbishment the foreman of the building company was using a disc cutter to remove some bricks at intervals from an existing wall so that a new brick wall could be tied into the apertures. Somehow, pressure must have been brought to bear on the disc which shattered into many pieces, some of which cut deeply into this man's forearm and a few other places. He was bleeding profusely and we wrapped his arm in a towel and I drove him to Southmead Hospital. It was not until I was driving back to the abattoir that I realised there was a pool of blood slopping about in the passenger foot well of my car, and how lucky he had been to get to the hospital so quickly as he could possibly have bled out. Fortunately, this man made a good recovery, albeit he was badly scarred.

In the days before health and safety regulations, people did what they could to keep the job going, and it was not unusual for one or more of the slaughtermen to climb a ladder up into the roof of the slaughterhouse to return a hawser to its pulley wheel or to release a hook which had got stuck on the rail. On this particular day the sheep gang foreman, Ken Dicker, ascended a ladder to correct some malfunction, leaving someone to stand on the base of the ladder to stop it slipping on the floor. The job took rather longer than expected, and the man at the foot of the ladder got fed up with waiting and moved away to do something else. Having sorted out the problem, the foreman started to come down the ladder without first checking to see if his anchorman was still at his post. Predictably, the ladder slipped with Ken's leg between two rungs snapping his shin bone like a carrot. Again, Ken made a good recovery.

On another occasion a man called Wyndham Fudge, while freeing a wire cable (part of the workings of a cattle stunning trap), managed to leave his middle finger behind the cable as the heavy door dropped which stripped much of the flesh from the middle section of his digit. We all thought that his finger would have to be amputated at the first joint, but the plastic surgeon drew the remaining tissues together and then as it healed proceeded over a period to burn with acid the tissues from the top and bottom of his finger until he ended up with a very disfigured but partly functional finger. This was a treatment I have never heard of again since.

One of the two partners who owned the abattoir was so focused on making money that when one morning he accidentally walked into a star, gashing his head and nearly knocking himself unconscious, he refused to go to hospital in the ambulance which had been called in case he missed an opportunity to make a shilling. The ambulance personnel tried to insist that he went, whereupon he had to give them a leg of lamb each to go away.

One of the young cleaners working at the abattoir was not the sharpest knife in the box, and on several occasions would fall for the same prank, causing endless merriment for onlookers. He would be using the high pressure hose to clean the floor at some distance from the tap when someone would surreptitiously put a hand round the corner and turn the tap off. The unfortunate lad would look towards the tap, and seeing no-one, would peer down the end of the hose, whereupon the hand would reappear around the corner and turn the tap back on. The high pressure of water would nearly take his head off, but the poor lad never seemed to learn from his experience.

Another youngster, who was training to be a slaughterman, was always getting up to outrageous tricks. One of the funniest I remember was when he secreted himself in a large wheelbarrow, used for carrying out the sheep skins. He persuaded one of his friends to cover him with skins, leaving a tiny gap for him to look through, and then to wheel him near to where one of the slaughtermen was working; a man with little sense of humour, particularly when the joke was on him. The lad in the wheelbarrow would start to talk intermittently to this man, who would stop, look all around, and then thinking he had imagined it return to his work. After a short while the process would be repeated, with a pair of eyes peering out between the sheep skins to watch the effect, and to know when he had been rumbled, to give himself time to make a hasty exit before he was clouted.

This same young man had a reputation for outrageous and reckless behaviour. He used to travel to work on a scrambles-type motorbike, and one day while we were all working he went out to use the toilet, returning minutes later astride his motorbike with a blow-up doll strapped to the pillion seat. He drove through the slaughterhouse and out of the far door onto the loading bay where the meat lorries used to back in, and where there was a drop of some four feet onto a concrete yard, over the edge of which he drove just like a stunt rider, fortunately coming to no harm. He was also known to have driven down through the main streets of Bristol, similarly accompanied, no doubt much to the amusement, and in some cases consternation, of passers-by.

On another day there was a very aggressive large billy goat in the lairage, and during the lunch hour this same prankster decided to imitate a matador, going into the pen and side-stepping as the goat would charge him. On this occasion he did not escape unscathed, returning to the mess room with a broken arm. Some would say he deserved everything he got!

As with most companies (unless they are very lucky) a certain amount of pilfering took place. One old boy would stay behind at the end of the day, ostensibly to do some final trimming and tidying up of the carcasses. This commendable procedure had developed into something rather more, and a long sliver was taken off each bovine neck and collected in a separate bucket, to supply a small but lucrative dog meat business. The manager of one of the companies that killed at the abattoir used to often stay behind to do his book work, and when he sat at his desk in his office, which was on the second floor of an office block across the yard from the cooling hall, the top of his white trilby hat could just be seen through the window. Not wishing to be disturbed during his dog meat collection, the old boy would ask a fellow conspirator to stand in the doorway and watch the top of the trilby hat. However, one day the manager, who had been in the meat game all his life and knew all the dodges, propped his hat up with a stick and came around the back way, suddenly arriving by the side of the lookout. It was by now of course far too late to shout a warning, so the manager and the lookout stood side by side watching the dog meat purveyor collecting his ill-gotten gains. Nothing was ever said but the point had been well made.

On a Friday, many West Indians from St Paul's (a district of Bristol now remembered as the area where the Bristol race riots took place and, in happier times, the St Paul's carnival) would come up to Gordon Road to buy their weekend meat. This usually consisted of old ewes which, having been bought, would be cut up by the abattoir butchers by placing them on a wooden block and chopping them into little pieces with a cleaver, as opposed to being done properly with knife and saw. This clearly was acceptable to these customers, but I often wondered how they managed to eat the meat when it was so full of little splinters of bone.

It was not only among the lower echelons of staff that dishonesty would occur. The scales on which these ewes were weighed had a high hook on which to hang carcasses by the leg, and also a flat platform below. The weight registered would appear on a large glass-fronted dial at eye level behind the scale. The owner of the sheep would hang the back leg of one or more sheep on the high hook, pointing at the dial with his finger to draw the attention of the purchaser in that direction, but all the while keeping his foot firmly on the flat deck below. Having made their purchases the customers would set off, often on foot, transporting their meat in various receptacles, such as cardboard boxes, shopping trolleys and even prams and pushchairs. The more affluent would come in their vans and cars and would probably buy several carcasses at a time. I remember on one particular occasion the back doors of a white van were opened to reveal three women, about six children and a couple of spare wheels all crammed into the back. Into this already overcrowded environment were squeezed five or six ewe carcasses for home butchering. Clearly this was less than desirable from a hygiene point of view, but unless I had evidence of the meat being subsequently being offered for sale, there was nothing I could do about it.

CHAPTER 11

THE
END OF THE ROAD

LIFE CONTINUED HAPPILY for some years at Gordon Road. Les Mawditt and Teddy Howick retired and the post of senior meat inspector for Bristol became available. I applied for the position and was successful. The weight of responsibility now descended on my shoulders, but as I have mentioned, I had been well schooled. A further meat inspector by the name of John "Rocky" Harris was appointed to replace me, and we also had a meat inspector's assistant, which with Les Worden and Louis Thorn, made a team of five.

Most of the time we would have students working with us, gaining their practical experience and clocking up the number of hours they were required to spend at an abattoir. These would be vet students, mostly from Bristol University Veterinary College (Langford), Environmental Health Officer students doing a degree course upgraded from a diploma in public health, and meat inspector students. Most would be local people.

A lecture room had been established in the midst of the abattoir complex where the students would be taught by various doctors and professors. Veterinary students from Langford would arrive in a coach, and Environmental Health Officers and meat inspectors would come from the Bristol Technical College courses, later to become the University of the West of England.

My colleagues and I would save up various diseased specimens for these lectures, taking care when vet students were due not to put out any with large abscesses in, knowing full well that without exception vets felt honour bound to cut open these abscesses covering the tables and the surrounding floor with gallons of pus: although it was not part of our job, we would always clean up the room afterwards and dispose of the specimens. The senior lecturer was Dr Dudley Osborne, a man of incredible knowledge, and from whom I learnt a tremendous amount. On some days Dr Alan Wright, a specialist parasitologist, would attend. He always used to say "Please only save me parasitic specimens; I don't know much about pathology"; I am sure this was not true: he, like Dudley, was a master of his subject.

I had noticed over the years that those who failed the meat inspection section of the Environmental Health Officer course often did so on account of their practical examination as opposed to their theory. When I became senior inspector, in an effort to alleviate this situation, I asked the Health Committee if they thought it would be a good idea for me to have all the Bristol-employed students at the abattoir on Wednesday afternoons, which was usually quiet, to give them some concentrated tuition on practical meat inspection and also examination technique. At any one time there would be 12 Environmental Health Officer students in Bristol as in those days Bristol City Council would sponsor four students every year.

The Health Committee welcomed the idea with open arms and the sessions began, initially with just the 12 students. When the students returned to college and told their colleagues, who came from all around the country, about our sessions there was a clamour to come to "Dave's talks", and eventually my audience was 20 or 30 strong. Hitherto I had only ever given one lecture, which was when Bristol Technical College asked if I would do a talk on game law and recognition. I shall never forget how my saliva dried up; I was almost shaking with nerves. However, once I started doing my own sessions I soon forgot my nerves and thoroughly enjoyed imparting my knowledge to other people, something which I have done ever since. I lose all track of time and I am equally at home talking to just a few or as many as 500. The longest talk I have ever done without a break was six hours.

As a result of these sessions at the abattoir the senior lecturer on the Environmental Health Officers course at Bristol Technical College (having now become the University of the West of England) Martin Wood, asked me if I would do some talks for the University. This started my official career in teaching, of which more later.

My appointment as senior meat inspector was generally well accepted amongst my colleagues, with one exception: this was an individual who, although he had qualified a year later than me, strongly felt that one of us should not be ranked over the others, and in any case he had not himself applied for the position. He took his grievance very seriously and would spend his time finding problems, taking great delight in dragging me, sometimes physically by the arm (he was a very strong man), to see what he had found, and enquire what I was going to do about it.

I put up with this for a couple of weeks, but in the end it got to a point where I asked him to join me in the office for a chat. I asked him if he would have preferred an outsider to become senior, or that someone he knew and had worked with for a number of years now took the reins. I told him I was not prepared to tolerate his attitude any longer so the decision must be his as to where we went from here. He thought about this for a while, and although we never discussed the subject again, his whole attitude changed towards me, seeming to adopt the role of my personal body guard.

I remember one occasion involving a young man who had been employed as a cleaner at the plant and who was of a very aggressive disposition, particularly if he had been drinking. I had complained to the management about some cleanliness problem and he had seen fit to take the criticism personally. He confronted me in a particularly belligerent and foul-mouthed manner and, not wishing to get into a slanging match, I turned away from him, suggesting that we go together to the owner's office to sort out the problem. As I walked away I knew this man was probably going to strike me from behind, but my previously uncooperative colleague stepped between us and said to this young man "If you dare touch Dave I'll kill you !" This may sound a bit extreme, but in the tough environment in which I have spent my working life aggressive outbursts are not uncommon, but fortunately rarely result in actual physical violence.

The new member of the team, John Harris, had had a very varied career before his arrival at Gordon Road. He had worked as a meat inspector some years before down in Devon, but also had on two separate occasions been a pub landlord, and for a time had his own trawler fishing off the Devon coast. He had not been with us very long before his entrepreneurial instincts took over when he noticed that the stomachs from the vast numbers of sheep we were killing were being thrown away. Having obtained permission from the owner of the plant to have some of these stomachs, he would take them home and mince them with a hand-operated table mincer set up in his garage. He found there was a tremendous demand from the dog owners in the area for his frozen packs of minced green tripe. It was not long before John purchased an electric mincer to cope with the demand. Soon after this his garage was no longer big enough and a unit had to be rented, and a little later still staff had to be employed to do the work while John worked long hours with us at the abattoir. His business increased to such an extent that he ended up importing boxes of frozen tripe from Ireland as well as what he was able to obtain from local abattoirs, and he eventually sold the business to a local pet food company.

The premises at Gordon Road, situated as they were amidst suburban houses, were no longer suitable for a vastly expanding business with large lorries coming and going at all hours of the day and night. The holding capacity for both animals and meat was no longer sufficient. As a result Roger Hendy and Ken Hill decided to build a brand new abattoir on Roger Hendy's farm situated in the countryside north east of Bristol, outside the city boundary. Plans were drawn up and permission granted for the new venture. Sadly, at around this time, Roger Hendy was taken seriously ill with a disease which was expected to leave him incapacitated, but which in fact finally killed him. Before he was taken into hospital, Roger called me into his office and paid me a tremendous compliment by telling me that I would be leaving local authority employment and working for him and his partner Ken as their manager of their huge new plant. Thereafter we had several in-depth meetings; latterly these took place at what became his death-bed in hospital. Roger was a very special friend and one of those few men that you meet during your life for whom nobody had a bad word. On reflection I decided for all sorts of reasons, even though Roger's widow continued with the project, not to become part of it.

With the eventual closure of Gordon Road no abattoir would exist within the city limits, therefore there would be no job for me with Bristol City Council as a meat inspector. However, I did have some months while the new premises were being constructed to consider my options, during which time Bristol very kindly attempted to find me alternative employment at my salary level. Two of these possibilities come to mind. The first of these was a health and safety officer at Bristol Airport, and I was invited to spend a couple of days with an existing officer to see if I would be interested. Although the work could have been very stimulating, it was not really my cup of tea.

However, while I was at the airport I witnessed something very unusual. This involved a business set up very close by which purchased dairy breed bull calves (known as "Bobby" calves in the trade), which normally would be slaughtered at about 10 days old for "Bobby" veal. These calves were flown alive to various European destinations to be reared in the so-called veal crate system. One or more aeroplanes had been completely stripped of internal furniture for this purpose. A huge tarpaulin had been spread throughout the fuselage to collect all the urine and faeces, and collapsible pens constructed throughout with an alleyway up the middle. Because of the cigar shape of the fuselage there was only room for one calf in the pen in the tail area (like a rear gunner), followed by pens with two or three calves in, gradually increasing to pens for about 10 calves, right up to just behind the pilot's seat. They were lifted into the plane in crates elevated by a fork-lift loader.

The second possibility was a post as superintendent of the city mortuary. Thinking to myself that one dead body would be much like another, and heaven knows I had seen enough dead bodies in my career so far, I said I would have a look to see what the job entailed. At that time the mortuary and Coroner's Court were situated in an old building in the centre of Bristol called Quaker's Friars. I was accompanied on my visit by a senior clerk who did all the administration for the mortuary. He warned me before we went in that the two operatives employed there at the time were unaware of the reason for my visit, and in fact one of them had a relative who worked in a mortuary elsewhere and who was apparently interested in the vacant post.

When we knocked on the door it opened just a fraction and a head appeared round to enquire the purpose of our visit. The clerk having explained, the door was then opened fully and we stepped into an environment which could have been from Dickens' times. The all-pervading stench was diabolical, and although I have a very strong stomach from years of working in abattoirs (but where most of the smells are of recent origin), this was decomposition in the extreme, which almost made me want to retch. The operative who had started to take us on a guided tour was obviously oblivious to the aroma. He took us first into a large room where there were four or six marble mortuary slabs. Between each pair there was a toilet bowl with no seat. I asked the operative, and immediately wished I hadn't, what was the purpose of these toilet bowls, and he told me they were for the disposal of the stomach and gut contents. There were no bodies in that room because the work for that day had finished, but before we left the room the clerk nudged me and said "Have a look at the wiping cloths"; I glanced at the neatly folded cloths and saw that they were nappies retrieved from dead infants. Carefully lined up next to them on each table was a set of dissecting instruments and saws, with lovely mother-of-pearl handles; the sorts of things that would fetch a princely sum in today's antique market.

We moved on to the refrigeration area where about 10 bodies were stacked one above the other on roller trays covered in sheets, and only the tops of mostly grey heads were visible facing in my direction. These were people who, having died, required post-mortems to establish the cause of death, probably only because they had not visited their GP in the recent past.

We then moved on to the deep freeze department where three or four bodies on similar roller trays were being kept. These were apparently suspicious deaths of one sort or another, including possible murder victims. I can still see now the body of a young girl with very long blonde hair hanging down from the tray, with only a piece of material to cover her modesty. All of them were frozen solid while investigations took place.

Directly next door to the tea room used by the operatives was another doorless room with only a couple of mortuary slabs, an area I was told that was set aside for the "dirty bodies". One of the slabs was occupied with what looked like an Egyptian mummy wrapped in some sort of sacking material. I was told that this was the body of a merchant seaman retrieved from the Avonmouth docks complex, and which had apparently been in the water for some six months and was in such an advanced state of decomposition that sacking was necessary to hold the remains together. This cadaver was clearly the source of the stench which pervaded the whole building. I never knew the reason for this poor man ending up in the dock.

I was reminded of this horrendous sight when years later a policeman friend of mine recounted a similarly macabre incident. When he was young and working in Bristol he was summoned, together with his colleagues, to the sighting of an arm and part of a torso protruding from the mud underneath the famous Clifton suspension bridge. It was decided that the best way to retrieve the body was to lay some ladders on the mud, and my friend volunteered to crawl out across them. He seized the arm to try and release the body from the mud but unfortunately the limb came off in his hand, indicating that the body may have lain undiscovered for many months.

When I was at school in Bristol we regularly crossed the suspension bridge on foot as our playing fields were situated on the Somerset side. One day, when returning from a cricket match carrying bats, pads etc, we saw a man on the other side of the road climb the barrier and jump into space. We all rushed across the road to look down but there was no sign of him; he must have fallen out of our sight underneath the bridge. Many years later, when discussing this incident with a man working at Spears bacon factory, he told me that many years before whilst climbing on the rocks with his brother below the bridge on the Somerset side a body had slammed into the rocks by the side of them and bounced on down. He said they were within feet of being killed. By comparing the time-scale and the position on the bridge we decided that is was very likely to have been the same incident.

One of the inspectors working with me at Gordon Road told me that some years before, when driving on the Portway, which is the road below on the Bristol side of the bridge, a body had hit the road just in front of a lorry carrying steel girders. The driver of the lorry had jammed on his brakes in an attempt to avoid running over the body, dislodging some of his load, and a steel girder went straight through the windscreen of the car behind spearing and killing instantly the woman occupant.

On a slightly happier note, yet another friend of mine was driving across the bridge towards Somerset when he saw a man trying to climb the barrier, obviously with the intention of jumping. Leaving his car in the middle of the road my friend raced towards the man and rugby tackled his legs as he was about to go over. Wrestling him to the ground he sat on him, kicking and screaming, and then dragged him towards the ticket office until he was near enough to summon further help from the attendant, who called the police.

While I was still at school a jet fighter pilot flew his plane under the bridge from Bristol towards Avonmouth along the Avon Gorge. He was protesting against the disbanding of his squadron, but sadly as he banked to the left having cleared the bridge he was unable to rise fast enough and crashed into a wooded bank on the Somerset side. If only he had kept going straight he would have been safe. For some time the area was cordoned off while the debris was collected, but as soon as the cordon was lifted boys from my school were scouring the area for souvenirs from the shattered aircraft. One boy actually found the pilot's thumb which he took back to school and pickled in formalin. He exhibited this grisly relic to anyone who wished to see it, until one day he was caught by a master, and the thumb was confiscated, hopefully to be returned to the pilot's family.

But I digress. Although the job at the mortuary would have been very interesting it also involved dealing with the families of the deceased, many of whom would have been in an extremely distressed state. Doing this day in and day out could well have depressed me. The council believed that the employees in the mortuary had too close a relationship with certain firms of undertakers, resulting in recommendations to the bereaved relatives, no doubt for financial gain. The superintendent would be expected to stop this practice, which could have made life very difficult for a newcomer put in charge of an established clique. For these reasons I decided the job was not for me.

By now I had decided that I would stay with what I knew, and an ordinary meat inspector's job luckily became available in Somerset at about the right time. I applied for it and was successful. This resulted in a considerable drop in salary, but living as I do on a smallholding I felt I could make up the shortfall for my family with various ventures. And so it was now time to say goodbye to Bristol, which I did with much regret as I had been very happy working for this very considerate authority.

CHAPTER 12

PASTURES NEW

AT THE START of my employment in Somerset in 1981, the district council, Woodspring, looked after four abattoirs, and later on a deer farm. I was now employed not just as a meat inspector but as a meat inspector/technical officer. My base plant was an abattoir and shop complex on the outskirts of the Bristol dormitory town of Nailsea, the name being derived from the fact that although now inland the settlement used to be"at the end of the sea". The nearest the sea comes now is to a town called Clevedon on the bank of the Severn estuary. On the edge of Clevedon was a small abattoir run by one man with occasional assistance. There was a further abattoir behind a butchers' shop in Blagdon, near the famous Blagdon fishing lake, which doubles as a reservoir for Bristol. These three would now become my responsibility, but when they were quiet I could be asked to help out at the Weston-super-Mare plant which, like Gordon Road, had been a local authority abattoir but was subsequently sold to the FMC Fatstock Marketing Company, later to be sold again to another FMC, this time the Fresh Meat Company.

On my first day at Nailsea I was viewed with some suspicion by the owners of the family business, which had been started by the present owner's grandfather who farmed in the area. It was a thriving business, doing a certain amount of wholesaling, but also supplying the adjoining shop, which although situated some distance from the main shopping areas of Nailsea was (and still is) a very popular destination for the discerning shopper. It is a place where the joint for the weekend or the steak or chop for the evening meal can be discussed on a personal basis. Their home-cured bacon and cooked ham is second to none and people know it is produced on the premises and mostly sourced from local farms.

The family knew of my history in the big city, so inevitably there was a period when we were cautious in each other's company, but soon they realised that I was firm but fair, and a mutual respect and trust built up between us which continues to this day.

Life moved at a much slower pace than I was used to in Bristol, and after a period of readjustment once again I began to enjoy my work. Most of the daytime hours were spent at Nailsea, but as soon as I finished I would drive to Clevedon to inspect their daily kill. In those days the kill could proceed without necessitating the fulltime presence of an official like me. It was incumbent on the abattoir owner to make sure that the correlation of each animal and its offals was maintained until inspected.

Once or twice a week I would go to Blagdon where the kill was very small – maybe two cattle and 10 or a dozen lambs and no pigs. Some weeks, when things were fairly quiet at Nailsea, I would be asked to stand in at Weston-super-Mare abattoir and come back to Nailsea and Clevedon at lunch-time. There was a resident meat inspector at Weston whom I found very difficult to get on with. He knew that I had had a much more responsible job in Bristol, and seemed to find it necessary to try to order me around, probably as a result of an inferiority complex. He was the sort of man who would stop the kill so he could have his appointed lunch hour. But on the days when there was very little to do in the afternoon he would quite happily forgo his lunch hour to finish early: a typical trades union jobsworth.

The men working at the Weston abattoir were much the same sorts of guys as I had worked with in Bristol. Once the initial suspicion about a new inspector with a certain reputation of standing no nonsense and knowing his job had died down, I made some good friends, and there were some real characters amongst them. It was remarkable that whereas the boys at Gordon Road seemed to know all about the employees of the plants in the surrounding area, the boys at Weston seemed to have worked in isolation for most of their lives, with only one or two exceptions.

Weston killed quite a lot of young bulls, an aspect of the trade completely new to me. I had up to now only been used to steers, beef heifers and cull cows and bulls. These young bulls were a way of using dairy breed bull calves which hitherto had been killed at 10 days or a fortnight old after they had drunk all their mother's milk which would be full of colostrum and therefore unsaleable. This very young veal was known in the trade as "Bobby" veal and we used to kill hundreds of them in Bristol. Later on there was quite a demand for them to be exported to go into the European veal rearing units. Now some of them would be reared as bull beef and fed predominantly on rolled barley, a practice that continues to this day. Their growth rate is phenomenal, and when they are slaughtered at around 18 months old they have developed all the characteristics of male bovines, ie the heavy shoulders, huge neck muscles and wide skulls. The advantage for people who sell this type of meat is that it is extremely lean, putting on huge amounts of muscle rather than fat, living on a high quality diet in an intensive situation. Some heifers and steers are also reared in this way, but tend to become over-fat for the market.

While working at Weston I became involved in a process which I had been aware of before. The process was the production of something called "Proten beef". This procedure on live animals involved the injection of papaya juice into the jugular veins. Some of the abattoir staff were trained in the administration of this procedure. The training given by a member of the company supplying the juice was very perfunctory, and involved a bucket of hot water and a cloth to swab down the neck of each animal before attempts were made to find the jugular vein, by which time the animal would be climbing the wall. It involved the restraining of each animal in a crush. The head would then be roped and pulled back over the shoulder exposing the neck, and the pressure would make the jugular vein more available for injection. The animals would bellow in pain as the juice was injected and I even saw some that actually cried in agony. The reason for doing this to the live animal was to use the animal's own heart to pump the juice around the body. One day a little heifer was so stressed she fell in the crush and the rope tightened around her neck and she virtually hanged herself.

The animal would then be released and allowed to stand for 15–40 minutes to let the juice thoroughly circulate in the body. The animal would then be slaughtered to produce Proten beef. The liver in particular became so soft that the inspector had to reject it as the consistency became like blancmange. All the beef was beautifully tender to eat and the product had quite a following. The ability of the papaya fruit to tenderise meat became known by the natives of Papua New Guinea when they wrapped their meat in papaya leaves to cook it in hot ashes. The knowledge became commercialised, particularly in South Africa where the process was used extensively to tenderise big plough oxen, which would inevitably eat very tough by virtue of their long hard lives working on farms. As far as I know the process probably still goes on out there today. The young beef cattle at Weston abattoir would have been perfectly acceptable for food without putting them through the sort of agony involved.

Instinctively I knew that what was being done was wrong, and how it had been allowed in the first place I shall never know. I had some of the slaughtermen at Weston come to me to tell me how cruel they thought the process was. It was difficult for me to do anything because

I was only the second man at Weston, and the resident inspector did not seem inclined to bother to do anything. Nevertheless, I went to see the chief Environmental Health Officer to express my concerns. He said that making a fuss would probably cause all sorts of legal problems and a resultant loss of business to the abattoir. I said that that did not concern me and added that if the slaughtermen were telling me that the process was "bloody cruel" I was not on my own in my concerns. He agreed to back me and the local authority mounted a national campaign to stop this process in this country. Inevitably when you spoil a money-making venture you tend not to be the most popular of people, and although the boys at Weston remained friends, the head cattle buyer whose idea it had apparently been would no longer speak to me and various threats against me were made. The campaign was even discussed on radio on the Jimmy Young show.

In the end the process was banned by Europe, but not for the right reasons. It was banned because the beef was considered to be not of the nature and substance expected by the general public, not because it was a cruel practice. It seems there are no depths to which the human being will not sink when it comes to causing extreme suffering to animals, whether it be for financial gain or for some religious dictat.

The senior slaughterman at Nailsea, David Hearn, was very much of the old school, and so we had much in common. He was very well known and respected in the trade, and very skilled at what he did. He was familiar, as I was, with many of the lovely old traditions of the trade, and was one of the few still able to back set and dress a spring lamb around Easter time, which was a special process for display in a butcher's shop window. At Christmas fat stock show cattle, which had been manicured to perfection for the show ring, would be dressed in the normal way, but the skin at the end of the tail with its plume of hair would be left attached, and this was also used for display purposes in the shop window. When the skin was removed from these animals, often Christmas trees would be carved with a knife into the sub-cutaneous muscle of the flank and brisket. In my very early days, when calves were allowed to be inflated with compressed air to plump up the meat of these very young animals, sometimes after evisceration the skin would be eased from the majority of the carcass but left attached down the back. A paper ruff would be placed around the neck and the whole would be placed in the shop window for display.

Can you imagine the outcry there would be if these practices were carried out today? Issues would be raised from a "public health" point of view and many members of the general public would be horrified because they have become so divorced from nature that they think meat is made in a factory somewhere and put on supermarket shelves in plastic packets. Nobody died as a result of the now old-fashioned way of doing things, in fact people were probably healthier then because of their natural resistance which they acquired from childhood. David Hearn epitomised "the good old days" of the meat trade, before bureaucratic madness descended on it. Very often today I find the trade is governed by people who know little of the history and reality of both the meat and farming industries. I have also noticed that the more incompetent these people are at the lower levels the more likely they are to be elevated up the promotion ladder where their stupidity has more power behind it. This is probably true of many industries today. I just happen to know all about the meat industry.

David and I became great friends, and he was a great support to me when I was readjusting to the slower pace of life in my new job. We worked together for many years, and shared common interests in the shooting and gundog world. In later years he stepped back from actual slaughtering and took on a more managerial and administrative role. His dry sense of humour was unrivalled, and he had the ability to walk down through the slaughter hall dropping a word here and a couple of words there resulting in all the men going at each other like bantam cockerels, when he would just walk on quietly chuckling to himself. Tragically, David died from a burst aortic aneurysm (he may have survived if the ambulance had reached him within the Government set response times).

Sadly, David died just when he was looking forward to very well earned retirement. He is still sadly missed in the workplace and beyond, and enriched the lives of all who knew him.

The abattoir premises at Nailsea were very old and really needed updating. The town was expanding rapidly, and as I had found in Bristol, it is not really desirable for abattoirs to be situated in residential areas. A parcel of land was purchased outside the town, between Nailsea and Clevedon. This particular site was chosen because the main sewer from Nailsea out to the Severn estuary at Clevedon ran right through it. Plans were drawn up and work commenced. The building was to be constructed on what had been the beach thousands of years before. This gave a firm foundation on a stratum of rock and soil, as opposed to conditions close by which were characterised by many feet of peat; in fact the northern end of the Somerset Levels. Building was completed and the move made in 1988.

The small abattoir at Clevedon was run by a man who followed his father into the business, and who was an extremely skilled slaughterman. He primarily killed animals for the local butcher's shops, all of which have now sadly disappeared. He did, however, have one client who would scour the country for old worn-out dairy cows, which if they passed inspection, which was not always the case, would go to some far-off boning plant to be made into reconstituted products. These were probably made to taste nice, however, if people were to see the sorry state of the animals concerned they might opt for fish and chips or even turn vegetarian, or just possibly they might go and buy some decent meat in the form of a steak or a chop so they could see what they were actually eating.

Again, the Clevedon abattoir was a very old building which had become surrounded by houses, and eventually the business closed and the site was due to be built on. The owner moved on to work in another plant in Somerset. Shortly afterwards demolition started and

I managed to purchase the old flagstone floor and many of the faced corner stones from the old building. The flagstones are now paths around my house and many of the cornerstones are incorporated in various buildings on my property. I did not pay very much money for the stones but in the end it was a very expensive exercise. At the time I was driving a large Triumph estate with an automatic gearbox. Overloading both car and trailer because the stones had to be removed as the building was being demolished around my ears I managed to wreck the gearbox which cost me £500 to replace, which in those days was a lot of money.

The little Blagdon abattoir was the second from last in the area for a butcher's shop to have a slaughterhouse behind it. This was very common in yesteryear, and this one was not dissimilar to the one belonging to my uncle in Neyland in Pembrokeshire, where I killed my first sheep at the age of seven. There is one still existing in this area, and how he keeps going I don't know because the owners of Blagdon (two brothers) found that with the advent of veterinary ante-mortem inspection, to pay for a vet to look at the animals alive, and then pay again for a meat inspector to look at them dead, was rather more than a small business could stand. They therefore decided to close down the slaughtering side of the business as it was cheaper to buy in carcasses from elsewhere. Also at that time the influence of supermarkets was increasing, which had an effect on the volume of trade in the shop. People commuting from the village to Bristol, Weston-super-Mare or further afield would probably do their shopping in the larger conurbations.

The elder brother tended to run the shop while the younger had busied himself with the slaughtering side and all the behind- the-scenes preparation. When the decision to stop slaughtering was made the younger brother found employment with the now very famous Thatcher's Somerset cider, whose headquarters were only a few miles from the village of Blagdon: another new start.

CHAPTER 13

DEER FARMING

IN THE EARLY 1980's when I came to work at Nailsea, deer farming was in its infancy, and there was no legal requirement for farmed deer to be examined by an inspector from the local authority. The venison market was then (and still is now) supplied by stalked wild deer and also farmed deer. The owner of this particular deer farm, who was one of the first in the business, approached Woodspring District Council to see if it would be possible for a meat inspector to examine his carcasses at his own expense and put a health mark upon them if appropriate. Woodspring decided that this was in order, and I was asked to visit when the owner let me know he had killed one or sometimes two deer. He also wanted some guidance concerning the refurbishment of an old stone outbuilding as a dressing and cutting room, complete with refrigeration and washing facilities. The deer were actually shot and bled out in the field and then carried to the dressing point on the back of a tractor. The venison was destined to be sold at the farm gate and occasionally at farmers' markets.

I tried to give this gentleman the benefit of my experience as far as the building was concerned, but it was evident from the start that this man was very meticulous about everything that he did, and hygiene and cleanliness were second nature to him. I remember on one particular occasion I told him that I could not pass one of his carcasses as fit for food because the bladder was still inside. This is something that happens occasionally, particularly in sheep, as it can easily be overlooked. The owner was absolutely mortified that he could have missed it, however the bladder was very soon removed and the otherwise immaculately dressed carcass was passed as fit.

Deer farmers like to point out that their product is slaughtered at the optimum time for young and tender venison. Stalked deer, on the other hand, can be of indeterminate age, and older animals tend to be less succulent and tender.

One of the owner's sons, while still at school, expressed interest in becoming a veterinary surgeon. Obviously he was very familiar with the killing and dressing of deer, but he asked if it would be possible to spend some time with me at the Nailsea abattoir to witness the killing and dressing of other food species. I think he thoroughly enjoyed his time with us at Nailsea, and later on I was to see him again at Langford Veterinary College as a budding young vet. Interestingly, his particular year of students was the one selected by the BBC for their programme Vet School and the follow- up Vets in Practice.

A few years later it became legally necessary to have the carcasses of farmed deer examined and health marked, and the premises in which they were dressed to be inspected and passed. It was gratifying to hear that when these particular premises were visited by a vet from the Meat Hygiene Service, which by then had taken over from the local authority on such matters, he could find no fault; however, he had to drop a few percentage points on his score, to allow "room for improvement".

By contrast, the rules are very different for stalked deer, many of which are shot through the chest, the reason being that if they do not die instantaneously by the time they have had a bullet passed through the chest cavity they are not likely to run very far. If they are shot a little far back, the bullet may have punctured part of the stomach allowing the escape of stomach contents into both thoracic and abdominal cavities. These deer, wherever in their bodies they happened to be shot, are gralloched (eviscerated) out in the wood or field, or maybe up on the hill. However well or carefully this is done, inevitably the carcass is very likely to be internally contaminated as it is dragged or part carried back to a vehicle. One can imagine the contamination which is likely to occur from hair, earth, leaves, grass etc, as well as if the stomach has been perforated, the leakage of stomach content. Many stalkers will carry a bottle of water to swill out the cavities after evisceration.

Can the powers-that-be not see the anomaly that exists between, on the one hand, an animal which has been killed and dressed in a licensed premises where every knife cut is bureaucratically supervised, and on the other hand an animal which is killed in its natural surroundings and eviscerated at the place of its death, which is something mankind has been doing for millennia with no apparent ill effect? The resulting meat from both these deer could very possibly end up on the same menu in the same restaurant.

As far as I am aware there has never been a food poisoning outbreak attributable to the consumption of game meat. The same situation exists between wild boar shot in the wood and wild boar which are farmed and slaughtered in an abattoir.

Stalkers today are advised to undergo several levels of stalking courses which test species recognition, the ability to shoot straight at the correct spot, and a rudimentary ability to examine the viscera removed from a carcass for any abnormality. The primary reason for this, in most cases cursory, examination is to check for the presence of bovine or possibly avian tuberculosis. Unfortunately, in most cases this is performed by superficially trained amateurs, although amongst the ranks of deer stalkers there must be some meat inspectors and vets who by virtue of their long training and experience would hopefully do a more professional job. Once again the contrast can be seen between the ad hoc inspection given to stalked deer and the rigorous examination afforded to deer inspected in licensed premises.

The vast majority of deer carcasses will, I am sure, be free of disease. However, some years ago I examined a roe deer carcass which had been killed and gralloched on the Mendip Hills, and brought to me by an extremely experienced stalker friend, purely because he did not like the look of what he found within the carcass and around the green and red viscera which he had sadly left where the deer had been shot. This was long before anybody thought of deer stalking training courses, and that deer might be harbourers of bovine TB. Although I only had the carcass to examine, I suspected the presence of TB by virtue of the fact that I found what are known as tubercular grapes both in the thoracic cavity and in the abdominal cavity. I told my friend whom to speak to in what was then the Gloucester MAFF office, to where the carcass was eventually taken. This was necessary because TB is a notifiable disease and suspicion of its presence must be reported to the ministry. This applies to several other diseases and parasites including anthrax and foot and mouth disease, and the larval stages of the Warble flies. Tests were performed and in due course my suspicions were confirmed that the animal was suffering from Bovine tuberculosis; this was the first time it was ever verified in a roe deer in this area.

CHAPTER 14

NAILSEA

BEFORE THE MOVE to the new abattoir at Nailsea in 1988, the old premises were my base for about seven years.

David Hearn was the foreman of the slaughtering gang and working with him were Mark, Stuart and Nigel. Between them they got through an incredible amount of work, and I would help out where I could, something which is forbidden today. In the early days of my association with the business it was run by the eldest of four brothers, who lived next to the abattoir and also farmed some adjacent land. The father of the brothers, together with two of his sons, would frequent the livestock markets, purchasing stock from as far away as the middle of Wales. Many lambs were bought when the price was right and grazed in the fields at the back of the abattoir, a reservoir to be drawn on as and when required. Many of these lambs were not familiar with the drainage ditches, some covered in duckweed, which are often used to divide fields on the Somerset levels. Some of the lambs would fall into the water, necessitating David having to walk these ditches every day to rescue the survivors.

The uncle of the family was a butcher, specialising in the curing of bacon and ham for which the shop in Nailsea was renowned, as is still the case today. He taught me the art of how to burn the hair off a pig, traditionally done on farms in the absence of sufficient scalding hot water. This is a skill not known to many people, and involved the sprinkling of barley straw on and around the pig in the sitting up position and then subsequently lying it on its back propped up by a brick on either side. This was best done in an alleyway with a through draft, and then the straw would be set alight. The secret was the thickness of the sprinkled straw on different areas of the body so that the burning would not cause damage to the skin or body but remove all the hair.

Later on, the eldest brother decided to go farming in Herefordshire and the third brother then picked up the reins and runs the business to this day, with the capable assistance of his two sons.

While I was still working in the old premises, I started to get requests for meat inspector, environmental health and veterinary students to spend time with me when down from college to do their allotted hours to be spent at an abattoir. The owners of the plant were very accommodating to me in this direction, bearing in mind there is no obligation whatsoever on them to allow this. This happy situation continues to the present day.

There were not many weeks when I did not have students of one sort or another with me, and it certainly kept me on my toes. It was very interesting to meet people from a variety of backgrounds, and I believe the abattoir staff enjoyed having different people around to talk to. I always limited the numbers at any one time, because space was at a premium and I didn't want their presence to impede the through-put. There were also safety issues to be borne in mind (which we were perfectly aware of even before Big Brother decided that everything had to have a risk assessment).

Nailsea specialised in slaughtering top quality stock, both for sale in their own shop and to supply butcher's shops, which were far more numerous in those days before the cut-throat buying powers of the supermarkets killed many off. Their beef was predominantly from black Hereford heifers. They were the by-product of the diary industry, by using a pure red Hereford bull on Friesian cows. The resultant calves had white faces, good conformation and flavour from their father and less fat from their mother, who was continually having to be put back in calf to produce her next lactation. The best lactating cows in the herd would be mated to a Friesian bull to hopefully produce female calves for replacements in the milking herds. The male Friesian calves were of little value. Many were killed as "Bobby veal" and only some would be reared for beef.

Pigs and sheep killed at Nailsea were the best available, except I do remember one occasion when the owner asked my opinion on killing some old ewes for the ethnic trade, a business that I was very familiar with in Bristol. I suggested that as these were unknown customers he should not give credit on the transactions, but otherwise to go ahead as long as the animals were properly electrically stunned but bled by a Muslim, making them acceptable as halal. Some weeks passed with a happy mutually beneficial business arrangement. Completely forgetting my advice, credit was then allowed and a large bill for both the ewes and the killing was run up, and the rest, as they say is history.

CHAPTER 15

NAILSEA: THE NEW ABATTOIR

THE PLANNING PERMISSION limited the size of the new abattoir at Nailsea, and as a result sadly compromised the internal layout. This has been a problem ever since. The slaughter hall was designed around what was known in Belgium, where it was designed, as a cattle-dressing module. This arrangement of lifts, platforms and saws took up very little room and was apparently capable of dressing eight cattle an hour, the idea being that the headless body went in at one end and two sides of beef came out at the other. Whether the fault lay with the original installers (which I suspect) or with the module itself nobody seemed to know, but whatever the reason, it never ever worked. The decision was taken by the owners, because time was of the essence, to dismantle most of the module and put together a short and less-than-ideal line system.

Because the revamped cattle line took up more room than the module would have done the sheep and pig lines became very congested, and space to move continued to be a problem. In spite of all the teething problems work began, and the ministry vet, who at the time would do a six monthly visit, observed that it was the jewel in the crown of all the abattoirs he visited in his area. He did, however, add to his observation by telling the owner that he wished he had contacted him for advice at an earlier point in the design and construction process. The vet knew that there were things in the legislative pipeline which possibly could have been beneficial to a newly designed line.

I now settled down into the routine for what was to be the next 24 years (maybe not quite finished yet). Gradually the staff changed and people came back into my life whom I had worked with before at Gordon Road and Hotwells. Namely Peter Worle my old friend from Hotwells days, and his stepson Nicky Jeffries, and Mike Rowley, who had when we were young taken delight in winding me up so much. Sadly, Mike later developed a brain tumour and died.. Also Barry Shiner, whom I have known since he was a boy when his dad used to drive his own meat lorry out of Gordon Road delivering all around Bristol. Ricky Hill, who was a lad at Gordon Road, joined us from an abattoir at Bath. John Mikalic, a very skilled butcher from Gordon Road, and Nigel Hawley, who was one of the originals from the old Nailsea slaughterhouse also joined. Harry Roberts, who had worked as a young man at the old Nailsea plant, came to us at Gordon Road and went on to their new plant at Westerleigh and then finally came back to join us. Harry, real name Peter Roberts, but renamed by us for a bit of fun after the famous murderer of three policemen who then went on the run, living rough in woodland for many weeks before being caught.

In 2011 a farmer called Ian Windell, who farms on the famous Badminton Estate in Gloucestershire, came into the abattoir to deliver a couple of cattle to be used in his on-farm butchery business. He was a man that I would go and have a chat with, and on this occasion the conversation turned to yesteryear and Ian said he knew an old retired slaughterman. I enquired as to his name and the reply was Fred Williams. My readers may remember that I recounted some of the names of slaughtermen with whom I had worked 48 years before at Hotwells. Among those names were father and son Arthur and Fred Williams. Arthur was the man who invented the first oscillating chine saw with the aid of a motorbike engine, which as I now know was supplied to him by his son Fred. Fred was the man who looked after many of our cars at that time.

Ian said he would bring Fred down to meet me again together with Pete Worle. In due course Fred came down with Ian and it was lovely meeting up again after all those years. Fred said he would like to arrange a reunion for all that we could find left of the Hotwells boys. He said that he would organise this on the next farm to Ian's on the Badminton Estate, where Fred's' daughter is married to the farmer: in due course, such as were left of us met over a sumptuous roast beef lunch with all the trimmings. There were in fact only five of us seated around the table together with a few of Fred's friends from elsewhere. What a lovely day we had, but it was sobering to think that as far as anyone of us knew no one else was left alive from Fred's time at Hotwells. I was able to let Fred have a photograph printed in the *Bristol Evening Post* of his father behind his invented saw which Fred did not have in his collection.

Back at Nailsea, initially I worked on my own, with holiday relief provided by Environmental Health Officers from Woodspring District Council. My holidays were only ever taken one day at a time, during the winter months. In fact I only took the job so that I was able to do this. This arrangement suited the Health Department very well because they only had to cover for me for single days. Whereas most Environmental Health Officers at the time quite enjoyed doing a bit of meat inspection, as a break from their regular work, some however really hated it, so one day was not too great a hardship. The reason for these days off concerned my main hobby. I train cocker spaniels to work on two large shoots near Taunton in Somerset. Fortunately my wife Alison, who has her own dogs, also enjoys these days away. It also gives us a break from the mini zoo and smallholding where we live.

The kill at Nailsea at that time numbered between 90 and a 100 cattle, 300 to 350 sheep and 200 to 250 pigs, with the occasional goat per week. All this I was perfectly capable of coping with, together with all the necessary paperwork of the day, concerning rejected offal and carcasses, and numbers killed. Today, with fewer numbers other than possibly goats and the odd buffalo, it is deemed necessary to employ two full-time inspectors and a vet. All that has changed in the intervening years is the amount of bureaucratic paperwork, which incidentally does not even cover the wide possibilities of disease and conditions which used to be interesting to record. Take, for instance, any problem with the kidneys of an animal: there are many possibilities to record. In the name of dumbing down and trivialising the whole job, possibly to accommodate the new regime of inspectors who do not know all that they should about the job, all that is necessary to record is "kidney lesions". What is that meant to convey I would ask?

The first few years at the Nailsea plant went by very happily as far as I was concerned. The job was made more interesting by my excursions to the little abattoir at Blagdon and the deer farm. Sadly, the little abattoir at Clevedon had already closed down. On occasion, when Nailsea was quiet, I would be seconded to Weston-super-Mare abattoir to help out with holiday or sickness times. Sometimes I would start early at Weston to get the job going and then come back to Nailsea to catch up on what they had done without me.

Weston then closed down and three of their employees found work at Nailsea: the foreman of the slaughtering gang, Dave Lintern, now retired but still a friend, Rob Stretch and Phil Broadway, both of whom had worked in the boning room at Weston, now brought their skills to Nailsea. Rob has since retired and Phil emigrated to Australia.

CHAPTER 16

EUROPE – BUT JUST THE BEGINNING!

MY WORK WAS now reduced to holiday and sickness relief for the two Nailsea inspectors, this usually only involved a day or two or maybe a week at a time. I am very lucky to have been able to wind down very slowly; so many people are working full time one day and no work at all when they retire. For those whose whole life revolves around their work this is very often a shock to the system, if they have little in their lives outside of work they often die quite soon afterwards. In my case I have always been in a position to have to find the time to go to work. In fact now that I am more or less retired, I often wonder how I found the time to do so at all.

The senior inspector for Nailsea and several other plants was Tony Mitchell. He has been a friend for many years and another to whom I taught the fascination of meat inspection. He knew that a phone call on an evening would see me standing in for some unforeseen event, like sudden sickness, by seven o'clock the following morning. This worked very well for a number of years, but one day the Meat Hygiene Service decided that they should put some more money into the coffers of their contracted veterinary agencies and demoted old casual meat inspectors such as myself with loads of experience to be the last to be called for emergency relief cover for holiday or sickness involving the regular staff. I was told that the veterinary agencies must be given the first opportunity for work; taking no account of my loyal service over the years.

Tony had been the senior fire officer in a fire station at Southmead, Bristol. He left the fire service through ill health, which has now improved. Looking for something else to do, he asked me as a friend what exactly I did for a living. I gave him a cursory account of what I did and he asked me if I would show him. I was due to be teaching on a course at Langford before long, so I told him to come with me on one of the days. I found him a white coat and some boots and stood him at the back of a group of students. He listened carefully throughout the day and at the end came to me and said, "I would love to do that".

I knew that the second meat inspector's course of two run at Langford was soon to start. Needless to say Tony was among the eighteen or so students enrolled on the course. I admire older men who embark on something like this, because there is so much to learn and so many new words to remember. It is a daunting prospect for a youngster, but a man in his middle years inevitably finds it hard going. It has to be said that most inspectors properly taught, as they were at Langford are men with previous careers behind them and for all sorts of reasons are looking for a change of direction. It is easier if they have come from the meat industry as many do, but a similar number have never been in an abattoir in their lives. Tony also had a farming background which stood him in good stead when it came to familiarity with farm animals. Like the others on the course he found it hard going but passed well, as did all the others.

I told him that I used to smoke many years before but had given up when I eventually saw the light. I smoked small cigars Castella Number 5's and every morning as I was one of the first in the local newsagents my paper and 20-25 cigars were placed ready on the counter. My smoking time was very limited being unable to smoke on the job, but all addicts find time for a fix. But the next morning as ever my daily ration was on the counter. After recurring bouts of bronchitis I made the sensible decision, and I put the last part used packet on the floor of the abattoir canteen and screwed my boot around on it. Tony smoked, and I hate to see my friends slowly killing themselves. I suggested that it would be a good opportunity with the start of a new career to start a smoke free life as well. He said he needed the drug for his piece of mind during his study time, but would give up at the end of the course. This did not actually happen immediately, but I am pleased to say he is now a non-smoker.

Tony rose quickly through the ranks helped I am sure by his abilities in his previous career. He was made a senior officer very soon and we now reach the anomaly that Tony became my boss because as a casual inspector, I was now fairly low in the pecking order. Tony would often when confronted by a problem at work bounce it off me, knowing that I had over the years encountered most of the problems. Tony made an excellent senior and in fact was asked to act up to the position of area manager to cover for maternity leave. He had the ability to defuse what could become an unpleasant situation. Later on yet another agency the FSA absorbed the MHS and decided to reorganise yet again and Tony somehow was left out of the loop. These office bound managers cannot recognise a good sensible man when they see one, but of course common sense is not the flavour of the month today, bulls..t and ingratiation seem to be far more acceptable,

Crazy bureaucracy was now biting hard into the meat industry, but of course it makes loads of work for the thousands of non producing pen pushers out there. Most of the madness came via the young European foreign vets now in charge. These generally pleasant young people with a predominance of girls are terrified of the large veterinary agencies they work for. There are obviously hundreds of others out in Europe and elsewhere waiting for the chance to come and work in this country. They must not be confused with our home-trained young animal doctors who hopefully go on to become budding James and Jill Herriots. Most established veterinary practices that do what vets are meant to do, which is to heal sickly animals, will not employ these young Europeans but a few of them end up working for DEFRA. Many OV's whom I have spoken to on the subject tell me that they would love to treat sick animals, but the glass ceiling is firmly in place.

On one of my holiday relief working days at Nailsea one of the first people I met on my arrival was a young lady OV. She was really pleased to see me, as she had been on one of the OV courses at Langford where I had taught her. With all the hundreds and hundreds I have taught over the years I am very good at remembering faces, but very bad at remembering names. Of course having done the various courses, the students remember my name which puts me at an immediate disadvantage. "Hello Dave" she said "How nice to see you!" I explained that I now only worked odd days and was not employed very often. We had a chat about where she had worked since her course and she herself was still only working as a relief vet for her agency, before hopefully getting her own plant somewhere in the country.

During the morning I took a break from inspection and looking out into the entrance yard, I saw a cattle lorry driver I had not seen for some time and went over for a chat and to catch up on recent news. I was fully aware that new rules insisted on me wearing a brown coat over my white overalls, when going outside to protect me from the contaminating fresh air. It was a bitterly cold winter morning in 2011. I had not been talking to my friend for very long before the young vet spotted me. She came running over and said, "You must have a brown coat, you must have a brown coat, in her slightly broken English. Deliberately misunderstanding her, I looked around us at the cold morning and said that I was perfectly warm enough thank you, "No, no, no. "She said "You must have a brown coat". I said "If it pleases you I will go and get one". Later in the day I thought I would have a bit more fun at her expense. I went over to her and said "That was very kind of you this morning, to be concerned about an old man catching a chill on a cold morning." "No, no, no." she said "You have to wear a brown coat when you are outside".

It is interesting that brown coats are now replaced by green since the FSA took over, "Let's have all new protective clothing never mind the expense." Only white boots should be worn in the abattoir and only green or brown boots in the lairage and outside. Surely 'clean' is the important thing not the colour. It is also interesting that never in all my years have I seen a cow, a sheep, or a pig put on different attire when coming from outside or from the lairage into the abattoir. They are still covered in faeces, earth, hair and whatever else, the same as they ever were. The really clever people who polish their trousers behind desks have no idea about abattoirs but they have the power to come up with all this nonsense to justify their unnecessary existence. Keep it clean, make sure it is not diseased and cook it properly. The outside of every chop, every steak and every joint, gets exposed to considerable heat with proper cooking. Not so reconstituted meats, such as burgers and sausages, which may not have sufficient heat applied to sterilise them.

A few years ago the Ministry vet on his six monthly visit, told me that new legislation required that I would have to wear a hairnet, and in my case a beard-net as well. The next time he came on one of his visits I had obtained a green net onion bag, which I had put over my head tucked inside my collar and replaced my safety helmet on top. As he approached me with a friendly greeting I turned to face him, and he visibly recoiled at the sight of me covered completely with the onion bag. I told him, that when the law required for us to be completely covered in, I would be first in line. He said "You've gone mad", I said "Not as mad as you." Again the cattle come in complete with hair on their hides; the sheep come in with wool covering their skin, pigs come in with stiff hair covering their bodies. All this animal hair has accompanying faeces, earth, grass and various other extraneous material. I wash my hair and body every day maybe several times, to keep contaminating material to a minimum. When it comes to abattoirs these hairnets are just a joke. It could be said, by the time these contaminated animals 'with no hairnets' had become meat there is a good reason to stop human hair from dropping on it, but consider the ultimate destination of the aforesaid meat, possibly to be prepared via a celebrity chef on the television for the consumption of some lucky consumers. There is no sign of a hair or beard net there! Haven't the priorities become somewhat blurred by the stupid know-nothing people sitting behind their European desks, justifying their unnecessary existence. I even heard that they wanted trawlermen pulling their nets in a force nine gale to wear hairnets. I would love to stuff their hairnets up their fundamental orifices.

What about the stalker or the man ferreting rabbits or shooting squirrels for the more unusual table. I have never seen a hairnet in this area. It must not be forgotten that all the aforesaid may well end up in the same restaurant and on the same table as the lamb, pork or beef so heavily regulated. To my knowledge there has been no detrimental effect from these practices. We are now making rules for the sake of making rules certainly not for reasons of common sense.

On two separate occasions pigs were delivered to Nailsea by private producers, (private producers are now a large part of the small abattoir business). They are probably not familiar with the procedure of ante-mortem inspection by virtue of the fact that they only come very occasionally to kill one or two pigs from their hobby farms. On both occasions two pigs were involved and the owners concerned trying to be helpful put their pigs straight into the stunning pen. When it came time to kill these pigs the slaughtermen assumed (one should never assume anything) that the vet had ante-mortemed the pigs (two different vets on two different occasions). The slaughtermen killed the pigs on both occasions. Bear in mind that at this particular small plant there is little else for the vet to do, apart from peering at their computers for hours on end. Because the plant is small a fulltime lairage attendant is not possible for financial reasons.

Very soon the mistake was realised and instead of common sense, madness took over and on both occasions the pigs were thrown in the bin. Obviously the owners on both occasions were hopping mad and in my opinion rightly so. The pigs were post-mortemed on both occasions and found to be perfectly fit for human food by the very experienced meat inspectors. Because the young vets were terrified to use their initiatives in both cases; working for a veterinary agency who would think nothing about sacking them for a legal transgression instead of commending them for using their heads, those pigs died for absolutely nothing. All the time and money spent by the owners to bring them to a point of rare breed perfection for food, totally wasted. Only a relatively few years ago the inspector only went to the abattoir to do post-mortem inspection. There is no doubt that a look at the live animals is beneficial as a precursor to post-mortem and something I always did of my own volition at Gordon Road and later at Nailsea. I did this early in the morning accompanied at Gordon Road by the very experienced fulltime stockman called Oggles. All the animals considered not to be fully fit, for all sorts of reasons, were categorised and kept until the end of the kill to give the inspectors more time for their inspection and if there happened to be an animal with some nasty potentially contaminating condition it could be kept separate from the carcases around it.

All that is needed for ante-mortem inspection is a good stockman; he is able to pick out a sickly individual from among dozens of its peers. Very few vets working in abattoirs in my experience have this instinctive ability. See the chapter about the arrogant Spanish vet who managed to pass a dying animal with septicaemia as fit to kill. This resulted in the animal being slaughtered in the middle of the kill, inevitably contaminating carcases either side of it.

If common sense instead of madness had triumphed on the days concerned wasted life and wasted protein could have been saved, by careful inspection by highly experienced officers who decided the animals were perfectly healthy.

Europe inflicted the OVS's (latterly the OV's) upon the British meat industry by saying that it was essential for only vets to do ante-mortem inspection. The fact that highly experienced highly knowledgeable meat inspectors had for years been doing the job perfectly well, counted for nothing to the Eurocrats. I believe that in other European countries they do not have professional meat inspectors the way we do and so, possibly in ignorance, Europe has replaced a very good meat inspection system in this country with a much poorer one and a much more expensive one. Personally, I have ante-mortemed more animals than many young European vets have ever seen, including the diagnosis of an animal with BSE referred to in another chapter. On even that occasion, an English vet tried to claim that he had found it, even though he was seven miles away at the time, and the animal would without a doubt have gone into the food chain, if I had not seen it.

One night a couple of years ago I had a phone call at about eight o'clock in the evening from the owner of Nailsea to ask me a favour. An Aberdeen Angus steer had been delivered amongst other cattle to be slaughtered the following day. The steer had leapt from the back of lorry-gang plank, slipped on the floor and snapped its lower leg. The favour was, could I come and help slaughter and dress this unfortunate animal. The reason for this was that the owner knew my ability to slaughter and dress food animals and also I was living closer to the abattoir than any of the employed slaughtermen, other than Nigel who lived in Clevedon and was already on his way. During my seven mile journey, I remembered what had happened to the pigs which had not been ante-mortemed by a vet. When I arrived the abattoir owner and his sons were already there, together with the owner of the animal. I told them that the position may have changed and that I must ring the young vet resident at the time. I asked her if with my experience could I carry on and slaughter the animal having ante-mortemed millions of food animals in my time. She said that she trusted me implicitly, but should anyone find out that she had not done the ante-mortem on this animal she would be sacked by her company. She said I could carry on but the animal would be thrown in the bin. She then said she would drive as fast as she could from Bristol to look at the animal. We had two options; the first to slaughter the animal on welfare grounds, or wait for half to three quarters of an hour for the vet to arrive. We decided on the latter option as the animal was lying down quietly, rather than trying to stand. What madness that an animal has to suffer for longer than necessary just to satisfy the bloody Eurocrats. It is absolute madness that a man who is entrusted to teach these young vets and had looked at more food animals dead and alive than they will probably ever see is no longer allowed to assess an animal's fitness for slaughter.

How much longer must we bow down to the European disaster? We have done nothing but go to war with most of the European countries over the years and millions of good men died to keep this country free and great. They must be turning in their graves to see what is happening with the connivance of our self-enhancing political class

CHAPTER 17

THE EU INVADES NAILSEA

NEEDLESS TO SAY, there was no help to come from the government then or since, because Europe are now the masters and we have to jump to their tune like it or not.

We voted in this country for a common market with our European neighbours, something I think most in this country would go along with. However, in the intervening years this has developed into a European state which, to my knowledge, nobody in this country at least has had a chance to vote for. I well remember a television programme I saw called I think "The Poisoned Chalice". In the programme they interviewed ex-Prime Minister, "shuffle shoulders" Ted Heath. When the interviewer said, "Mr Heath, the people of this country voted for a Common Market and this has developed into a United States of Europe, which nobody was told about or given the opportunity to vote for," Heath, grinning, responded that they "the political class" had known what was to come. "Ah yes, Mr Heath," said the interviewer, "but you did not tell the people that".

I think most of the current problems in this country can be traced back to our joining the EC. Thankfully, we did not join the Euro, but even that has not saved us from contributing to the various bailouts for countries unable to control their own finances. We will never again see the billions squandered in this way; we ourselves are still in debt thanks to the Labour party's spending spree with our money. They even had the audacity to leave a note when ousted from office saying that all the money was gone.

The people who did this to our economy should be called to account for their irresponsible behaviour. But, as with everything else they do, the political class slither away on a trail of slime having made their personal fortunes off the backs of the rest of us. When will the world turn, and the people of this country finally have some decent, honest human beings to vote into a position of power before it is too late?

Coming back to the problem of small abattoirs, surely it was not beyond someone's wit to leave them alone to supply local communities with their requirements, usually slaughtered in a much kinder environment than is inevitably the case in today's huge meat factories where quantity and through-put have to be the order of the day. By nature of their business, these large plants need to export large parts of their kill, which necessitates conforming to all the dictats of Europe. In this country we had a good system in place which served the country well for several generations. My suggestion, too late now because most of the small plants have gone, would be for home consumption to carry on as before, with a designated Home Health Mark. Every large plant would have an export stamp but would have to pay the penalty of vets-in-charge with all the additional costs involved, but this could be absorbed much better over huge numbers of animals.

Inevitably, as Europe has tightened its grip, we had to have a vet-in-charge at Nailsea. The first one to appear was an English vet who had a small animal practice in Bristol as well. He could not resist the temptation to offer his services as an official veterinary surgeon (OVS) to the local authority of Woodspring, now North Somerset, receiving at the time somewhere in the region of £60 an hour purely to ante-mortem the animals prior to slaughter, and to deal with any carcasses I might detain for his decision on rejection or not. Thinking back, what a kick in the teeth that was for me: I was earning only a fraction of his figure but I had ante-mortemed more food animals than this man had ever seen. Also, I had rejected thousands and thousands of pounds worth of meat and meat products over the years. All of a sudden I was no longer capable of doing the job because Europe had said so. Instead, they sent in vets, who in many cases hardly knew one end of an animal from another and thought they had a better system.

I began to play what I called "the vet game" with this man, and have played it many times since. I would detain a diseased carcass for his inspection. He would appear and closely examine the said carcass, walking around it and occasionally prodding or poking it. I knew full well that he did not have a clue what he was looking at. He would eventually say to me "What would you do?" I would respond that maybe the whole carcass could be rejected or possibly a leg or a shoulder which might be harbouring an abscess. When I told him of my assessment he would say "Oh yes, that is what we will do then". The bloody man was living off my back!

The next official veterinary surgeon was a different kettle of fish altogether. John David was a well-known lecturer on genetics at Langford, now retired and helping out his pension. He was a tall Welshman with a presence, a voice like Richard Burton and a very likeable disposition. We became good friends during his time working with us. He knew that I was on top of my job and never tried to pull rank on me; if anything, the reverse was true. He used to do everything he could to help me and the slaughtermen keep the job going. He would even help to remove some of the spinal cords from the cattle (specified risk material for BSE reasons). I can see him now climbing up some steps and using a pointed spoon that I had sharpened for him to make a nice clean job. John finally retired and has since died, but will be fondly remembered.

I played the vet game later with a vet at Nailsea, who was also a nice person; but this did not stop me from showing these vets how little they knew about meat inspection. On this particular day we had a lovely little heifer with a very rare condition which I have only seen about five times in my working life, and considering the millions of cattle I have inspected in 50 years of meat inspection, my reader can see how rare this condition is. The condition, known as eosenophilic myositis, is generally systemic and necessitates the rejection of the entire carcass. The symptom is iridescent green patches in the musculature usually throughout the carcass. The animal has usually done well and appears to suffer no ill effects from the condition. In all probability there would be no problem from eating the flesh from such an animal, but rejection would be on the grounds of being aesthetically repugnant; in other words, one would rather not eat something like this from choice. As I have said already, I have always worked on the principle that if I wouldn't eat it, or give it to my family to eat, then in nobody is going to eat it. This criterion has I think served me well over the years.

This vet was working at Langford at the time and has since taken up a post in New Zealand; as I said, he was a pleasant man. What then transpired came all the more as a shock for me. Playing the vet game, I showed him the offending carcass; he walked around it and looked at the green patches showing in the musculature. After some consideration he said "Perhaps we could mince it." "Mince it?" said I. I could not believe what I had heard. I then asked "Have you children at school (he was a comparatively young man)?" "Yes I have," he said. I said "In that case would you like them to eat this in their cottage pie?" "Of course not," he said. I said "Well in that case it must be rejected; why should anybody else's children have to eat it?" "Do you know, I hadn't thought of that," he said, to which I replied, "*That* is the job".

At about this time the responsibility for meat inspection was taken away from local authorities and a new government agency came into being, the Meat Hygiene Service.

A succession of mostly young European vets employed by one of the several veterinary practice-based agencies operating at the time now emerged. Since then, these agencies have bought one another out, and now there are only a few left with a stranglehold on the meat industry.

One of the early ones supplying OVS's was a company based in Exeter. I got to know several of the senior partners in this practice, who were purely in it for the incredible amounts of money to be earned for very little effort. I remember well when newly qualified vets in Spain, Italy, Portugal and Eastern Europe were coming out of the universities. The partners of this firm would go out to these countries literally (as I saw on one occasion) rubbing their hands in anticipation of recruiting more naive young vets to bring over here to earn the company a fortune. OVS's have now become official veterinarians, OV's, to comply with the European system.

Some years later I became involved when teaching some of these mostly very nice young people at Langford Veterinary College, and practically at several abattoirs, to work as meat inspectors initially. Later, if they were considered good enough, they would do a further three week course on which I also taught; then they could work as OV's.

These young people are used as fodder for the European system. They do not get a fair crack of the whip. The veterinary companies, who exploit their qualification for monetary gain, do not know whether or not they are proficient at the job, and still less it seems do they care. I do not care who does the job, as long as it is done well and the youngsters are properly monitored and maybe mentored for six months before being let loose on the meat trade. I consider myself very lucky to have done the job when I did and the training and mentoring I received was second to none. But imagine going to a strange country, with a strange language and indigenous meat inspectors (such as are left) keeping you at arm's length because of resentment at improperly trained foreigners taking their jobs. No meat inspector training is presently done in these particular European countries. The veterinary qualification in these various countries is considered sufficient to carry out the job. *Take it from me, it just isn't!* Some of the horrendous stories coming back to me of diseases and conditions missed by these young people make me cringe, and I have seen many failings with my own eyes. We were doing OK in this country but we had to be railroaded into the EU and just look where variously that has got us.

Like all these agencies, the Meat Hygiene Service was top heavy at the start with loads of taxpayers' cash thrown at them, and later on, when they have squandered that, they have to pull their financial horns in under the guise of *reorganisation.*

At the start of the agency, local authority inspectors were automatically taken over. I was but one of 300 odd who said NO. Some like me went for redundancy, not liking the sound of the phrase "harmonising the job". Harmony is a nice word; it is pleasant and relaxing. However, in this context, it meant grind the bastards down to the lowest common denominator. About three weeks before finishing, I was contacted by a man called Steve Frampton who it appeared I had known when he was a youngster, as a result of his exceptional prowess as a clay pigeon shot. Now it appeared that, as a grown man, he had joined the administrative part of the MHS. "Dave," he said "we are desperate; we do not have enough inspectors and we have to hit the ground running on Monday morning." He then asked "Would you consider working for the agency as a casual inspector?" I said "I think I have made it fairly clear that I do not want to," but continued "Make me an offer I can't refuse", which he did and I accepted. This situation still applies today. Hitherto I was going to start my own cut stone wall business, a skill I have taught myself and which I find very relaxing and satisfying. I think it has a certain permanence about it, something I have missed doing what I do. I had a break of service on the Friday and started for the MHS on the Monday doing exactly what I had been doing before.

There used to be weekends held up and down the country at expensive venues where we were wined and dined, and told what a good job the MHS was doing. I believe they call it team building. I have never heard such a load of nonsense. I remember at one of these weekends they divided us up into several teams; remember we are talking about professional men doing a very responsible job. We were told that we had to roll up some newspapers into long paper snakes and walk around the venue carrying the snake between us to see if we could work as a team. If it had not been so pathetic it would have been ludicrous. Is it any wonder the country is in the state it is in when we have idiots like those running the courses and those above them, who think this sort of thing is a good idea?

Initially nothing really changed. I had a visitation from the Local Area Resource Manager, who had been a meat inspector previously and someone who I had known for a while. He had in fact asked me if I would, amongst others join him in a private agency, hiring ourselves out to the M.H.S. At the time the idea did not take off, unlike today when the large veterinary agencies hire large numbers of staff out to the Food Standards Agency. This man asked me about my paperwork system at Nailsea as far as it went. He said "Does it work?" I said "Of course it works". All you need to know is how many of each category of species you have killed, what was rejected and why, and how long did it take you to do it. What else do you really need to know?" In those days at the end of each month a form would be filled out categorising all the different diseases and parasites found, this could be useful nationally for statistical purposes. Today there are hundreds of forms, but when it come to a particular disease of perhaps the liver or kidneys all that is necessary to record is liver lesions or kidney lesions. As with so much else today everything is being dumbed down, be it teaching children at school, the quality of University degrees, or the accurate recording of food animal disease, all in the name of keeping the political statistics good.

The next innovation was the recruitment of technical assistants who had to deal with the burden of bureaucratic paperwork, and also interesting other work such as counting sheep's teeth for BSE reasons. This is something I have always found a bit obscure, bearing in mind there has never been a sheep with BSE, other than the ones having the prion injected into their brains under laboratory conditions. Sheep have their own encephalopathy which is called scrapie. The reason for counting the teeth was to see if any adult teeth had risen, which happens around 12 months of age, but again this is a very inexact science. The appearance of just one tooth makes the animal "overage". This has a serious bearing on what is rejected as specified risk material. In an "overage", the head, the spleen, the ileum (which is the last part of the small intestine) and the spinal cord are rejected, which necessitates the splitting of the spine of the sheep to remove the cord. In a young animal, only the spleen and the ileum are considered to be toxic waste.

Now there were three officials – non productioners – watching five slaughtermen –productioners – working at Nailsea. I know that it is necessary to have an element of officialdom in the meat industry, but previously one watching five had seemed a reasonable ratio and was easily within my capability.

Today they have got rid of the technicians and replaced them with meat inspectors. Some of the technicians upgraded to meat inspectors but that was in the days when we used to train indigenous people from our own country, to do a job for our own people, until Europe decided that that was not good enough.

It seems to me, Cameron, who specialises in running with the hare and the hounds, started by saying he was going to get rid of a lot of bureaucracy and regulation; instead of that he has increased it. Until he throws off the European yoke and we do our own thing for our own people it can only get worse. It seems that politicians of all callings (except UKIP) are besotted by and afraid of anything European. They forget that down through history we have fought and beaten most of the European countries and in some cases we saved them from the Nazi jackboot.

Millions died during the last two world wars to keep us free. They would now turn in their graves if they could see what they gave their lives to stop has been achieved through the back door by several generations of self-serving politicians. Some of these become MEP's to churn out acres of unbelievable regulation to justify their existence at a vast cost to the European taxpayer. I am beginning to lose hope that a leader will come out of the woodwork to save us, but all good men and true never get involved in politics. What can we do but vote for UKIP? At least that would get us away from the European tyrants.

So now Nailsea had to settle down to try and support not only their own employees but also the horrendous burden of three officials instead of one; but of course it does not end there because the various tiers of bureaucracy above also have to be maintained to satisfy Europe.

To increase business and so be able to support the excessive bureaucratic invasion, Nailsea turned to the game trade. There was a need for a facility to undertake the slaughtering and butchering of deer, from as far away as Cornwall, so with the guidance of Julian Ridge Nailsea put in the necessary equipment to meet the need. Julian Ridge had his own deer farm and slaughtering facilities and was very well known in the deer farming fraternity.

Deer are a completely different proposition to cattle, sheep, pigs and goats. They can jump very high; to accommodate this ability very high wooden panels made from plywood were installed to form an alleyway up to the cattle-stunning trap. The trap itself had to have a smaller inner trap fitted inside it with a trapdoor lid on the top through which the deer could be shot. All this had to be made easily collapsible to revert to killing cattle. Now came another problem: introducing the deer to the alleyway, having parted one from its peers. This was a major problem, but the next was even worse. Deer can charge you and they can kick you simultaneously with both back feet and, if all this fails, they can rise on their back feet and beat you around the head with the front feet. To protect ourselves from these defensive attacks we had to carry plywood shields, with handles on the back, to parry the various blows. The majority of these deer were young stags with fairly small but sharp antlers. Apart from the danger to us and the extreme stress to the animals, this worked after a fashion, but I kept thinking there must be a better way.

We had an ante-pen prior to the slaughter pen for sheep. On one side there was a seven foot wall on which I could stand, so I thought that if we drove five or six into the pen at a time we would not need to stress them by separating them. Unlike sheep, which all tend to look away from you, deer follow you wherever you go with their eyes. I stood on top of the wall with my pump action rifle firing short .22 bullets; I could drop all of them in less than a minute. They did not even look down at their fallen peers. It was a much better system, but for safety reasons the boys retired completely from the area. They returned as soon as I finished each batch, hoisting the deer on leg chains to bleed them. Out of interest I had the pen completely cleaned before we started so that I could scour the floor for bullet heads which may have exited the skulls of the deer. I never found one, but I did find bullet heads later, when I inspected the carcasses and offal; they were often lodged in the lower jaw or the tongue. This is of course the advantage of low velocity short .22 bullets. I used thousands and thousands later on in 2001 when killing sheep in the foot and mouth disaster. Interestingly, using the normal long .22 bullets fired through my long barrelled pistol I have found that young bulls, and more recently buffalos, are very cleanly dispatched in the same way. In fact Nailsea have had to have a further licence for killing buffalo and they obtained one on the understanding that I kill them with a free bullet. The high-ranking vet, who came to watch me before the licence was granted, said that she was very happy with the procedure. When involved in the foot and mouth massacres I met many different vets, most of whom expressed their satisfaction with the efficiency of our procedure back then.

We killed deer at Nailsea for some time, but there was a problem with communication with the Deer Society. It seemed that their delivery of the live animals was very efficient and worked well. Sadly, removal of the butchered boxed meat would hang fire with us longer than was good for it, and what started well came to an end purely through lack of organisation.

CHAPTER 18

LIAM FOX

BLAGDON WAS THE next to go, finding that they now had to pay for a vet to go from Langford Veterinary College to look at the very few animals, and then later in the day to pay for me to go *with travelling* to inspect the kill. How a tiny little business was meant to survive such an expensive bureaucratic onslaught, especially as the supermarkets were doing their best to put all small butchers out of business, I do not know.

I thought enough of it at the time to go and see my MP Dr Liam Fox about this tragedy. He seemed very sympathetic and I offered to take him to Nailsea and meet the owner face to face. He accepted my invitation and we had a very, as I thought, constructive chat about the whole issue of small abattoir survival with the onslaught of bureaucracy coming out of Europe. Dr Fox arranged for us to go and meet some of his colleagues in the lobbies in the Houses of Parliament. Worried about the future, the owner of Nailsea had formed a club of similarly minded owners and others of a like mind up and down the country.

Having been escorted through the crowd of protesting miners by the police, this being around the time of the 1984 miners' strikes, about 30 of us convened at the appointed hour in the lobbies. What a magnificent building that is; one could spend all day just admiring the architecture from yesteryear hailing from the time when Britain was Great. There to greet us were Liam Fox, David Nicolson and Tom King. We were invited to express our views and opinions, which appeared to be well accepted at the time.

Beware politicians, however; they may appear to be supportive of a good cause but this is only a facade. They already have their own agenda on party lines and Liam Fox, a comparatively new MP, must have been bucking the party line held by John Selwyn Gummer who was the minister of agriculture at the time.

It became obvious that this well-meaning upstart, Liam Fox, must have rattled a few cages on party unity because shortly afterwards he rang the owner at Nailsea to say that he could not help him anymore and immediately put the phone down.

After this incident Liam began to rise through the ranks of the party in opposition, ending up as shadow defence secretary, later in power to be the defence secretary. One can only speculate that, having had a slap on the wrist for daring to be a proper politician representing his own constituents who put him in power, he must have agreed not to kick over the traces again and become a "politician" and grovel to the party line.

I met him again some years later in his surgery at Clevedon. This time it concerned the ownership of hand guns. The government were bringing in legislation to ban the ownership of hand guns after the Dunblane murders in 1996; more kneejerk reaction to an emotive issue in the public eye at the time. The fact is that most if not all legally held hand guns were held by probably the most law-abiding members of our society, most of whom only want to drill holes in pieces of cardboard and would not entertain the thought of criminality or murder.

Similarly, common sense should prevail in the policing of the so-called knife culture. I think the world has gone mad when I hear of a police officer in Clevedon tell a friend of mine who lays carpets to remove his Stanley-type knife from his belt or else he would arrest him; I have to carry my knives in a closed box. Perhaps these people would like me to do my job using my teeth. In any case it is not the knife or the gun, it is the people behind them that are the problem, and if the intent is there, as it often is today, why not just use a brick, but do it quickly before they ban bricks as well.

The third time I met Liam Fox was 2006 when I was selling my first book at the North Somerset Agriculture Show. He came into the marquee on a meet-the-voters visit. As he approached my table I stepped out and took his hand firmly and said how nice it was to see him again. You could see in his eyes his lack of recognition, but he must meet thousands of people. Without letting go of his hand I asked him quietly when he was going to save us, because the country was bleeding and bleeding badly. He said "it was not as easy as that"; my rejoinder being that that was exactly what he had said the last time we had spoken. This time his eyes said "I must have met this madman before". He enquired as to what I was doing there (thinking I suppose I had escaped from somewhere). His voice by this time had risen a couple of octaves because I still had hold of his hand.

I told him I was selling a book on the foot and mouth outbreak in 2001 (another political debacle). He asked if he might have a look. Releasing his hand I escorted him to my table. He said he thought he would like one, to which I responded that if he had £12 in his pocket he could have a signed copy. He rummaged in his back pocket and produced the required sum, whereupon I asked if he would like it signed for Liam, or Dr Fox. He said Liam would be nice. He took his book in his hand and disappeared down the tent at great speed, having escaped the clutches of a madman.

CHAPTER 19

BSE CULLING

MY YEARS AT Nailsea rumbled on with the company of many OVs and several different technical officers. Life was pleasant enough, with only seven miles to travel to work. I also ran a small game farm in conjunction with keeping a small syndicate shoot just outside of Portishead, close to the M5 motorway.

With Weston-super-Mare abattoir and the small abattoir at Blagdon closing down, this left only Nailsea and the deer farm for me to inspect. While still working for Woodspring District Council, if the abattoirs were quiet I would be given other work to do, such as the inspection of horse-riding establishments, dog kennels (both breeding and boarding), and at that time there were quite a few butcher's shops still hanging on against the unstoppable march of the supermarkets. I met many interesting people while working in this manner. It was not dissimilar to when working as a young man in Bristol I used to do the piggery visits around the city.

After a number of further years working as a casual employee of the Meat Hygiene Service, it must have been decided from on high that two full-time inspectors were needed at Nailsea, so a young man whom I had taught a few years before was to be the lead inspector and I was to be the second man. With most individuals I am sure that I could have accepted the situation and been quite happy to have taken a back seat. But this particular inspector wanted to make a name for himself, and very soon had upset the happy atmosphere built up over many years: he was very much an officious official.

There were several incidents where I had to pour oil on troubled waters. These culminated one day when we were dressing some lovely spring lambs. There can occasionally be a problem when using electrical stunning, but it can also manifest itself when animals are shot with a captive bolt pistol. As mentioned before, the shock to the system can rupture some capillaries (minor blood vessels) around the body; in particular the diaphragm, heart and abdominal wall are affected, but the whole body can sometimes be involved. The time lapse between the stun and the stick is also a factor in how bad the problem can become. The condition is known as "blood splashing" and can occasionally be so bad that the carcass has to be rejected on aesthetic grounds, but this is rare. The muscle is full of red haemorrhages which are still visible after cooking. Local trimming is all that is normally required.

On the day concerned two of the very valuable spring lambs had this blood splashing subcutaneously on their backs. The condition can manifest itself in several different forms, sometimes big red blotches about as big as a pea, sometimes tiny little spots about as big as a pinprick. These two lambs were of the latter form which was, to the untrained eye, hardly noticeable. If these lambs' loins had been cut into chops there would have been no problem from the saleability point of view.

However, this young inspector, wishing to show how powerful he was, proceeded to knife off all the subcutaneous tissue from the backs of these lambs, leaving just the muscles showing. This made these very expensive animals totally unsaleable. His display of enthusiasm caused all the slaughtermen to stop work to watch this mutilation, and even the vet, a nice man called Chris, came and stood by me to watch, asking me what I thought he was doing. I could do nothing but shrug; apart from anything else it was not this inspector's job to cut other people's property without permission.

This amongst other incidents finished me, and although this young man was not typical, it showed me the way in which the job which I had loved for so many years was going down the road of bureaucratic madness, enforced by a high percentage of people who had little or no knowledge of the meat industry, but an in-depth knowledge of trivial bureaucracy and their power to enforce it.

I spoke to the area resource manager of the MHS (another friend), telling him it was time to hang up my knives. He said "Please do not do that Dave," so I told him I could no longer work with the sort of people now appearing as inspectors in the industry. This manager was of the old school, and had worked originally as a slaughterman but later as an inspector; he knew the game. He suggested that I might like to work on the Over 30 Month (OTM) scheme. I accepted the offer and worked on from 1998 to 2001. This was yet another scheme dreamt up by the politicians this time concerning BSE. The scheme ran from 1996 – 2005 and was put in place to allay public anxiety over BSE. The BSE myths had been pushed along, panicking the population, by people like Professor Lacey who said we might all die from CJD, now called variant CJD, and that almost every blade of grass was now infected in the country.

It was decided that all bovine animals over 30 months of age should be slaughtered and burnt as a precaution. Instead of just doing that by taking animals to pre-arranged sites, killing them and burning them (obviously too simple a solution for the clever people in power), they dreamed up the OTM scheme to be acted out in designated cattle abattoirs up and down the country. This meant that appropriate cattle would be delivered from farms, ante-mortemed in the usual way by a vet, then slaughtered and dressed in the normal manner – saving the hides. They would be inspected for fitness for human consumption and then, instead of applying a health mark, they would be slashed, stained bright blue, cut in manageable pieces and taken away for incineration in large bulker lorries. Imagine the cost of buying the cattle from the farmers in the first place, processing to the point of sale to the butcher, and then destroying everything except the hide (which could be salvaged).

You could not make it up! This is what the European and home-grown political classes did to us. It cost the British taxpayer billions – for what? Presently, so-called over-aged cattle in this country have their brain stems examined in a laboratory for signs of BSE. The carcasses and offal are detained until the all clear is given and then they are stamped and allowed into the food chain. This was a process carried out by France all the way through the BSE years without any apparent harmful effects.

Not this country, we have to go down the road of knee-jerk reaction (as we always do with everything), costing us a fortune.

Do not think by my apparent complacent attitude towards all this that I am not aware of the tragedy of the mostly young people afflicted by this terrible disease called CJD. I do in fact have my own theories about this cattle disease, which I will discuss later.

During my time working at this large-cull abattoir I saw many very interesting diseases and conditions. Inevitably, older animals will develop more problems than younger animals. For someone like myself who finds these problems of riveting interest, and being a known compulsive collector of these various specimens, I found myself in seventh heaven, albeit I considered the job itself a costly waste of time.

The senior inspector at this plant had over the years been aware of me and me of him. But we did not really meet until working together at this plant. He knew that I would go miles for a good specimen, and he very gently warned me that all the resultant meat from the "cull" belonged to the Intervention Board, or in fact the Government, and that I should not take any pieces for my students. Hearing him, but being totally unimpressed, I would secrete smaller items about my person until a break period, which happened when each inspector had done the circuit of "heads", "red offals", "green offals" and "carcasses". We then went on half-hour break periods, before starting the whole circuit again. On my own in the inspector's office, I would transfer the said items to my large lunch box having first eaten the original contents.

One day there was a visitation by the head of the Intervention Board regarding the cull. As is usual with these sorts of high-powered visitations there was an entourage of lesser bureaucrats surrounding the main man. The main man appeared to be a pleasant enough human being, so I decided to state my case. I left what I was doing and went over to where he was standing, watching the kill. I was covered in various liquid splashes, an occupational hazard in my job. I enquired as to whether he was the head man, to which he replied that he was (the expressions on the faces of the entourage were a picture): it was obviously not the done thing to speak to someone like him without an appointment. I said that in that case I would like a favour from him; he enquired as to what that might be. I explained that I did quite a lot of teaching of veterinary and other students, and in order to do so I used many morbid specimens, some "clean bone" and some pickled in formaldehyde. I told him that some of his intervention cattle suffered from various interesting conditions and I would like to remove some of these specimens to show to my students. He asked as to whether I could bring the students to the specimens, instead of removing them elsewhere, to which I explained that he obviously did not know much about this subject, because some of these conditions come along every hour, some every day, but some just once in a lifetime.

He said he now understood why I was asking the favour, but asked "Who are you anyway and where do you live?" I explained about my past life and how teaching was now a part of it. I told him the name of my local village and he said he had a brother who lived nearby. I said

"In that case we should stick together." He said "Ring my secretary on Monday morning and you will have a dispensation." This is normally only granted to official bodies or institutions, not private individuals.

In due course all the necessary paperwork arrived, but it required that I should fill out about five copies of a movement licence form, with the weight, and the ear tag number etc. When you consider all this had to be done for something weighing a few ounces, I could not be bothered and so continued to take and secrete the small items and only got involved with the paper chase for the larger items. The whole episode caused consternation to the senior inspector who could not believe my audacity to ask as I had. I have found in life when you want something, go to the top and don't mess about with the minions. The good news was that my collection, which numbers hundreds of bottles and bones etc, was greatly enriched by my time at this abattoir.

The incredible waste of potential food on the Over Thirty Months Scheme was staggering, when you think of how much money we give in aid to countries, where simple people are starving to death. Much of this largesse is filtered off by the criminal element in charge in most of these countries, and rarely gets to those for whom it is intended. Surely it is not beyond somebody's wit to have brain tested all this tonnage of beef and, if found to be OK, to have frozen it and dropped it to the starving people, many of whom are otherwise doomed to die. No! We would rather waste it than maybe rescue a few lives from misery. What sort of people are running this once great country of ours?

CHAPTER 20

BRIDGWATER

I WORKED ON the BSE cull for several years, from 1998 until 2001, when the abattoir closed down as an abattoir and the slaughtermen went out in teams of four to slaughter the foot and mouth victims on farms around Gloucestershire, including the Forest of Dean. I did a bit of intermittent inspection work at Nailsea and at a horse abattoir near the Chew Valley Lake, which was built as a reservoir to serve Bristol. This work was very sporadic.

I heard from various friends in the ministry at Gloucester that a new scheme had been dreamt up by the Labour Government of the day. It was to be called the Farm Animal Welfare Scheme (I think this title turned out to be one of the biggest misnomers I have ever heard in my working life). The cruelty I witnessed during my involvement with this scheme surpassed anything I had witnessed in all my years as an inspector.

The cull abattoir decided to get deeply involved with this new killing scheme. The site was redesignated as a killing point; one amongst a number of strategically situated abattoirs around the country. The slaughtermen were recalled to do the killing of animals which could not be sold in the normal way because of foot and mouth restrictions. Farmers were getting short of animal food and money, and the Government were buying these animals under various categories in order to kill and bury them in landfill sites, or in the case of over-age bovines, to be burnt as in the cull scheme.

I signed up with the ministry to be a supervising presence on the cattle lorries going all over the place, particularly in Wales, to collect thousands of head of stock to bring back to the killing point for slaughter and disposal. Unlike the cull, none of these animals was inspected. Many animals were not fit to travel due to being too old or too young, or suffering from some debilitating problem or other. Many were trampled to death in the lorries. All of these animals had been licensed as fit to travel by a vet. Only some vets went to the farm and put down those considered to be unfit to travel.

Because the slaughtermen had returned to the killing point from the killing fields, the ministry were very short of qualified slaughter personnel, so I teamed up with a long-time friend and fellow qualified slaughterman and we offered ourselves to the ministry, initially in Gloucestershire and later on around Taunton in Somerset. Our experiences over around a nine-month period were the subject of my first book, which I called *Through My Eyes*, because I witnessed so many sad sights and had farmers crying in my arms after we had killed all their cloven-footed animals more or less in their back gardens.

After this I returned to meat inspection when Prime Minister Blair decided to try and kill terrorists instead of cows after 9/11. The cull abattoir had been refurbished a bit after doing the welfare scheme and we returned to the madness of the Over 30 Month scheme. I found that with the late start at around nine o'clock I was having to leave home well in advance of the start time to be sure to be there, never knowing how much traffic was going to impede my progress. This meant waiting for maybe an hour for the kill to start.

There was another cull plant operating next to the old Bridgwater market, and I was offered a job there, and even though it was about four more miles for me to travel, there were no traffic problems. The Bridgwater boys were a really nice bunch of lads, as were the inspectors, some of whom I knew already, and who came from far and wide. The plant was smaller and the kill was around 200 to 300 a day. The tubercular reactors, from all of Devon and Cornwall, came to Bridgwater on Wednesdays, and they normally numbered 60 to 80. Vets and field officers from the ministry office in Exeter used to come to Bridgwater to collect a selection of glands from each reactor to see if their reaction to the TB test was correct.

While working in Bristol and examining the green offals, I had noticed that there were many cases of what is known as "traumatic reticulitis". This condition is an inflammation of the reticulum, the second of the four stomachs in a ruminant animal. The inflammation normally takes the form of an abscess or sometimes multiple abscesses brought about by a sharp foreign body penetrating the stomach wall and allowing the escape of stomach contents which are full of bacteria.

Most foreign bodies end up in the reticulum, even the drug and magnetic boluses placed in the animal with an applicator pushed down the oesophagus. The drug boluses are designed to slowly release wormers to control parasites. When exhausted, the boluses leave behind a solid lump of steel, which is a weight, to hold it in place in the stomach. These are usually highly polished from being tumbled in the stomach and remain in position for the duration of the animal's life. The magnetic boluses were designed to hopefully attract sharp metal objects to stop them from causing damage to the stomach.

I decided to closely examine each case of traumatic reticulitis to see what was causing the problem. There was of course a smattering of nails, staples, screws and pieces of fencing wire; but predominately I was finding pieces of very thin wire about two to three inches long and usually pointed at both ends where they had rusted through. I considered for some time from where this wire might be emanating. I suddenly realised it was pieces of reinforcing wire coming from the tyres used by farmers to hold down the protective plastic sheeting on the top of silage clamps. After some years the tyres perish, exposing the wire, and some of the tyres are damaged anyway. Pieces of wire rust off and rattle down into the silage underneath.

Many of the cull cows that came in with this condition were comparatively young animals, which had no doubt become unproductive because of the intense pain that they must have been suffering from. The farmers would have been totally unaware of their problem and the reason for it.

Having sorted out why the problem occurred, I went to one of my farming neighbours and removed all his damaged and perished tyres and took them up to the tip. He thought I had gone mad and I really do not think he believed me as to the reason why.

It was probably a year later that I read in a veterinary journal an article written by a ministry vet, working in Winchester. He was describing a condition he called "car tyre wire syndrome" (CTWS). I rang him up to congratulate him and tell him about my observations over several years. I asked him how many post-mortem examinations he would do on average in a 12-month period. He told me it must run in to tens of cattle. I told him that I did at least 200 to 400 a day and that the frequency of CTWS was in epidemic proportions. He was very interested to know the extent of the problem, and I told him that I could get him some frequency figures for his research. Sadly he never came back to me.

Life at Bridgwater was pleasant indeed. Work always flows better if the company is good; I carried on collecting specimens and occasionally using my derogation. At about this time I received a telephone call from the owner of a veterinary agency in Worcester. He obviously had heard about my extensive teaching over the years, and asked if I would be prepared to teach young European vets to become meat inspectors.

The idea quite appealed, and at least I would get the chance to teach them some common sense as well. Having many contacts in the meat industry I was able to use several facilities unavailable to most. The students arrived in groups of 16 or 18. Initially, I used the facilities of the post-mortem room and the little abattoir at Langford; showing them the vast extent of the subject in the post-mortem room, and taking them through the killing and inspection procedures in the abattoir. Most of these vets were very keen, pleasant young people, who were very appreciative of what I was able to offer them in the way of training.

Having got the initial idea of the job under their belts, the important thing for these vets was to be doing the job for real under experienced supervision. Some 16 or 18 of them at a time was more than what was acceptable at one time in any abattoir because they would get in the way of any line system. The last thing I wanted to do was to upset or cause resentment among my trade contacts, who have been very good and understanding over the years, allowing me to use their facilities as my own. The owner at Nailsea in particular had been very long suffering in this direction over many years, and several hundred students of all relevant callings have passed through this plant.

Returning to these particular young vets, having given them a basic insight, I broke up the group, sending a third down to Devon to a meat inspector contact down there, a third to a lecturer friend in Dorset, Eric Harvey, and I kept a third with me. The latter I would fluctuate between Nailsea, to show them high quality food animals, and Bridgwater, to show them much more cattle disease amongst the older cull cattle.

One day while working with a group at Bridgwater I had one of the most embarrassing experiences of my lecturing life. I would ask the inspectors and the slaughtermen to save anything of interest for me to show the students. This worked very well until one day the slaughtermen had saved me a cow's uterus with what appeared to be a mummified dead foetus inside. I described to the students the various ways that the uterus protects itself from infection, by either mummifying a dead foetus, or else reabsorbing the soft tissues, leaving just the bones inside the uterus. In this particular case, from external palpation, I could feel this calf-shaped object inside the uterus. The students were keen to see the foetus, so I made an incision in the side of the uterus and out popped a flat brown horse-shaped object about eight or so inches long. It is not often that I am lost for words, but to see what was ostensibly a dead foal appear from a cow's uterus left me speechless. The shock only lasted a few seconds, although it felt much longer, until I realised what must have happened. The slaughterhouse staff, unbeknown to me, had found a small child's plastic toy out in the yard, where it had been run over and flattened. They had carefully and with difficulty introduced it into the bovine uterus. They had been working as usual, but the whole time watching me with the students. When I had eventually opened the uterus, they fell about laughing. The significance of the wind-up was probably mostly lost on my foreign students, but I still have the horse in my collection today!

One day at Bridgwater a cow came in with a huge tumour in its eye. This is an uncommon condition usually found in white-faced cattle, which lack pigment in their faces and are susceptible to ultraviolet light, which can cause tumours. This one had had the tumour for a long time, and as it was summertime and the tumour was weeping and bleeding, it had attracted the attention of flies. This poor animal did not only have a tumour which had been left far too long and had grown to the size of an orange, but was also crawling with maggots.

In the good old days it used to be the responsibility of the senior meat inspector to bring prosecutions for all sorts of misdemeanours committed under the meat inspection rules and regulations of the day. This was something I did a couple of times in my career; both times for acts of gross cruelty. Today the sort of case I have just described is the responsibility of the vet at ante-mortem, but because the offence is committed outside the abattoir he or she has to refer the case to the Trading Standards Department of the local authority.

I waited to see what the vet would do in this instance. Nothing seemed to be the answer, so I forced the issue by asking him what he intended to do about it. Grudgingly, he decided to call in the Trading Standards Officer. This man looked at the head of the poor cow, and after some consideration he thought that a farm visit might be in order. I asked him why. He said that he could then assess the conditions on the farm. I said "This is the animal you need to be concerned about," but he just shrugged and left. This typifies the way the world is today: people who do not even know what they are looking at, but are in a position to do something about a case of blatant cruelty, usually do nothing. Here was definitely a case to answer not only in keeping this animal alive for far too long, but also to have travelled it to the abattoir where every time its eye came into contact with other cattle or solid objects it must have been in agony.

I do not believe in leaping to prosecution as a first resort, but in a case of blatant disregard of the law, particularly if cruelty is involved, I think a successful prosecution has a knock-on effect to others via the bush telegraph, and cases like the one I have just described hopefully become very rare indeed.

The cull was coming to an end and only cattle born before July 1996 were now eligible for the cull. Only a few abattoirs were retained to do the last few stragglers, nearly all of which would have been old suckler cows whose working lives would have been a lot longer than commercial milking cows. Bridgwater now decided to clean the place up a bit and embark on killing some young beef for human consumption again. It was a difficult market to get back into and turned out to be a non-starter: like so many others, Bridgwater closed down.

I now moved in to very occasional work at Nailsea, only working to cover holiday and sickness relief. I decided that as I was getting older I did not want to travel any longer, and as Nailsea was only a quarter of an hour away it suited me very well.

There were two inspectors working at Nailsea now, both of them well known to me. One of them, Steve Curry, when working with me a few years prior, turned to me one day and said "It's your fault I am doing this bloody job." I enquired as to how that could possibly be the case, especially as I had not taught him. He told me that as a young lad he had visited the abattoir (Gordon Road) with his father, who was a biology teacher coming to the abattoir to collect some "cows' eyes" for dissection in some of his classes. Steve told me that he had asked "this man" working there if he could see the animals in the lairage. Apparently "this man" took him on a guided tour of the whole plant. Steve's interest in the meat and slaughtering industry was awakened. On leaving school, he trained as a meat inspector and qualified on a course at Smithfield Meat Market in London. "Who do you think 'that man' was," asked Steve. Of course, I had no recollection of the incident, but apparently I have to take the responsibly for starting Steve on his chosen career!

Timmy Bissell, the other inspector at Nailsea, whom I have also known for quite a number of years, is probably one of the most affable men I have met as a colleague. He really should not have taken meat inspection as a career: he is far too nice a man and really hates to upset anybody by rejecting one of their animals. Of course, there are those occasions when things get a bit heavy and you have to lay down the law, which does not always please everybody. Tim is a fountain of jokes which sometimes I have great difficulty in understanding, but his real great charm is his uncanny ability to spill the vegetable ink used on the ink pads to stamp the carcasses with a health mark. While at Nailsea we had several hilarious episodes of his ink spilling speciality, but he really surpassed himself while working at a big pig plant, where purple-coloured ink was used for export purposes. He managed to pour some of the ink from a new bottle over the head of the senior inspector, some over himself and some on the floor; he then went to find a mop and spread the dye everywhere while trying to clear up the mess. You cannot help but love him!

CHAPTER 21

MISREPRESENTATION

COMING AS I did from a butchering background, I am as aware as anyone that the reconstitution of meat and offals into various small goods is possibly open to abuse by unscrupulous operators. It is very easy when manufacturing such products to incorporate all sorts of ingredients which, were the consumer to be aware of them prior to consumption, would probably put them off such products for life.

Products such as faggots were originally designed to use up parts of the animal which had a very short shelf-life, particularly in the days before refrigeration. Others such as black pudding or haggis made use of animal parts which might otherwise have been wasted. That was of course in the days when any sort of protein in our diet was of value, unlike today when the waste of food in general is considered to be the norm in "developed" countries. As I have noted before, this is obscene when much of the "developing" world is short of food of any sort.

When starting out on my working life in a butcher's shop in a Bristol suburb called Henleaze I was very familiar with the manufacturing of these various reconstituted food products. Faggots were very popular:

the ingredients comprising the liver, the heart, the lungs, some of the spleen and maybe a bit of pancreas to sweeten the mix. Each faggot would be wrapped in some of the omentum fat which encompasses the intestines. We were very proud of our faggots, and our sausages and the brawn we would make out of pickled pigs heads, maybe with the addition of a belly of pork or possibly a hand of pork (the lower half of the shoulder). Even back then items such as pigs' heads were not every customer's cup of tea, but when reconstituted into a coarsely minced brawn it sold very well. All this was on a very small scale, and butchers of old took great pride in not only their recognisable joints, steaks and chops but also in their reconstituted products.

There are a few surviving butchers today who are equally proud of their products, but the vast majority of pies, sausages, pasties and burgers etc available now are mass produced.

My first introduction to mass-produced products was probably when working at Spears bacon factory in the centre of Bristol. Inevitably, there would be many by-products manufactured in what was in those days quite a large meat factory and, with the exception of their black puddings, I was very happy to eat all of their delicious pies. I also enjoyed their sausages, including their Polony sausage, which was a way of using up all the broken pies damaged in the production process with the addition of all the small goods which had not been sold on the delivery rounds. All this would be minced together using Spears' own recipe and the resultant conglomeration would put inside a red plastic skin; it was in fact very tasty.

Spears' sausages used up all the less saleable items, including the old sows and boars which had finished their breeding lives and could be usefully incorporated in the sausage mix. It was an outlet for the pig farmers who regularly supplied bacon pigs to the factory. I always remember that when incorporating these old (and therefore tough) animals into the mix it was generally accepted that 20 parts of sow meat to one part of boar meat would be used. This was to mask the unpalatable boar taste associated with adult male pigs. In this particular case, I always thought it would be easier to bin these old worn-out boars, but there was a certain unwritten obligation to the pig farmers for the factory to take these animals just to get rid of them. Apparently they did not wish to waste them and were prepared to risk tainting their products by using them.

Later on at Gordon Road a dealer started bringing these old boars in for us to kill, maybe 10 or a dozen at a time. He had an outlet with a firm making salami-type sausages, and he was able to buy these very low grade (but very big) old animals for next to nothing in the markets. Obviously a good profit was to be made. The boar taint could be masked by the high level of seasoning in this type of sausage. I remember that these boars arrived at the abattoir placed nose to tail, with partitions in between them in the cattle lorry so that these mostly huge pigs could not kill each other. By the time they were unloaded some of them were so aggressive (foaming at the mouth) that they would attack almost anything that moved, which included the slaughterman who had the perilous job of trying to stun these formidable beasts. Some of these animals had heads and necks so wide that the electrolethaler (electric tongs) would hardly bridge the head. The slaughtermen tried to work together, and while one would try to apply the tongs across the brain, a second man would shoot the pig with a captive bolt pistol as soon as the animal was immobilised by the electricity. The whole process was very unsatisfactory, both from the welfare of the animal point of view, and that of the men: pigs attack from below, and I have seen the terrible damage to a man's leg having been bitten in the back of the thigh by one of these large creatures.

Knowing that I owned free bullet capability, I was asked if I would shoot these pigs one at a time in the stunning pen. There was a sliding metal door to the pen and I would open this just a little. When the pig became aware of my presence it would often charge in my direction, allowing me a clear brain shot as it hurtled towards me. Pigs are best shot lower down the forehead than one would shoot cattle. Many,

many years later, when involved in the foot and mouth slaughter, I became aware that other slaughter gangs who were shooting pigs with captive bolt pistols were not doing a very good job, having not shot pigs in their normal jobs. They were only used to electrocuting them and therefore did not know that it was best to shoot them low down the forehead. This was something I learnt as a boy with my uncle Billy Nash in his little private slaughterhouse in Neyland in Pembrokeshire. This held me in good stead when it came to shooting these large boars.

One of the inspectors working with me at Gordon Road subsequently took a job as a hygiene officer at a frozen faggot factory.. Shortly after he took up his new post at the factory he told me never to eat any of the faggots from this well-known manufacturer. As usually happens when small business becomes bigger and bigger, quality tends to suffer. This firm, which had enjoyed a good name for their faggots, were bought out by a much larger company. As previously described, faggots should be made out of pig offals, but to bulk out the products other ingredients had been included such as pigs' heads, which can just about be justified coming as they do from the same porcine source; but then "blue" cow flanks, which could be bought for next to nothing, became part of the mix together with various liver and onion essences. Meat described as "blue" meat was derived from the very poorest of old bulls and cows, which had been milked virtually to death prior to slaughter. Their flesh seemed to have what appears to be a bluish sheen upon it in place of any fat. These animals were only one step above condemnation for emaciation.

I shall never forget one Saturday morning about 40 years ago, when working at Gordon Road, a small consignment of cows arrived for slaughter. They were part of a contract by a company who, amongst many other things, made beef burgers. They were at the time, and still are, a household name. There were 14 individuals in the consignment, and when the back of the lorry was opened and the restraining fences were placed in position on either side of the tailgate these poor cows tried desperately to walk down the slope. When arriving at the ground level, several promptly collapsed with exhaustion. Two had already gone down in the lorry and had to be dragged off.

After they had all been put out of their misery and dressed out, Teddy Howick rejected 12 of the 14 cows, in various conditions ranging from parasitic emaciation through pleurisy and peritonitis to septicaemia. After the news had filtered back to this well-known company over the weekend, fairly early on Monday morning an extremely upset company representative arrived at Gordon Road. He confronted Teddy in his office, telling him that he could not do this and that this "blue" meat was exactly what they wanted to make their burgers. Looking at this apoplectic individual over his glasses, as though he had just crawled out from under a stone, Teddy told him in measured tones that he should be thoroughly ashamed of himself. At the time his company was running a TV advertisement for their mouth-watering burgers, showing two men looking very smart and clean in white coats and trilby hats looking along lines of best quality beef carcasses hanging up at probably Smithfield Market, pens and notebooks in hand, pointing at this or that carcass, leading the potential consumer of the product to believe that this was what went into the manufacture of their burgers. Teddy reminded this irate representative of the company of their current advertisement, and although this was probably in the days before the Trades Description Act came into force, the whole sorry episode was set up by big business to mislead the eventual consumer. Teddy told this by now fairly shaken individual that if he had the audacity to send further consignments of debilitated and suffering animals to Gordon Road again, he should expect a similar outcome to that which he had just experienced and that in fact Teddy would make it his business so to do. He then told him to remove himself from his presence, hopefully never to return.

Needless to say, this company did not kill anything ever again at Gordon Road. They were obviously trying to do their deceptive business elsewhere around the country where there were no Teddy Howick lookalikes. I did hear that they killed a bit at the old Nailsea slaughterhouse, but it did not last.

I have witnessed in my own career the complete absence of any sense of wrong-doing amongst these large companies. It has to be said that the business is so huge, and the bulk of ingredients probably pass through many hands before ending up at the factories actually making the reconstituted product: there is no accountability. In fact, younger people have never known what the real product should taste like, unless they visit one of the few remaining small butchers or producers who still take a great pride in their pies, sausages, burgers and faggots. Much of the huge obesity problem which has taken so long for the public to address with mostly little success can be laid at the door of the fast food giants.

Young families today (many broken, if ever properly formed) have little time for the sit-down family meals of yesteryear when traditional roasts and stews and chops and steaks were the order of the day. Children not only learnt about what they were eating but, through mealtime conversation, about what was happening in the wider world. They learnt how to converse with adults and it was time well spent in the growing up process.

Today people are in such a rush to keep up; youngsters are given some mass-produced rubbish on a plate and told to go and watch the television while they eat it. Is it any wonder that so many of them are barely able to hold a conversation with an adult, let alone an in-depth discussion? Sadly, many of the last couple of generations are now reduced to monosyllabic low life, by being let down by an educational system which has been so badly tinkered with over decades. In many places, particularly in the more overstretched and deprived areas, it turns out young people who are barely able to read or write, or probably worse have little sense of right or wrong. This is by virtue of the liberal-minded elite who banned corporal punishment some years ago, so that the rest of society has to endure at the very least bad manners and at worst armies of vicious, callous, criminally minded sexually orientated thugs.

Not surprisingly, citizens of the calibre I have just described will happily eat any old rubbish put in front of them, and have done so since childhood. I do wonder about the rest who consume these products without a care. Surely the price alone must give them a clue that they are not eating prime rump steak when they tuck into their burgers and lasagnes. My own son, who I thought I had indoctrinated sufficiently on the subject of reconstituted meat products, came home one night munching on a burger in a bun. I told him that he could partake of any meat he desired in a recognisable form – chops, steaks, and joints etc – "But please do not eat anything reconstituted." He smiled and said "But it tastes nice." "Of course, that is the secret of selling crap, make it taste nice!" I replied.

It came as no surprise to me when the news broke in Ireland that "big time" criminals had moved into the huge fast food market. I also found it very interesting that it took some Irish scientists to isolate horse and pig DNA in burger meat. Our own sanctimonious, nit-picking Food Standards Agency should have been there probably years ago to protect our public health, but they have been doing their best to play catch-up ever since, and I think these revelations have plenty of mileage in them yet.

There are appeals in the newspapers all the time for donations to try to stop the horrendous cruelty involved in the transportation of horses over many miles across Europe to their slaughter destination. These sensitive and terrified animals are treated with such contempt, and their flesh is of such little value, that the lesser parts end up in the cheapest possible reconstituted meat products.

Horse flesh has the potential problem of possibly containing phenylbutazone ("bute"), which is a pain killing drug commonly used by vets (and possibly horse owners) amongst our indigenous horse population. I suspect that the vast majority of horses whose flesh has been used in our reconstituted meat products, not only in this country but right across Europe (this scandal is a big one), are coming from poor eastern European countries where veterinary care is minimal, and therefore the use of drugs of any sort probably is extremely limited amongst farm livestock. However, virtually all the horses slaughtered in this country are shipped abroad for human consumption where the perception of eating horse flesh is no different to that of eating the flesh of any other ruminant or non-ruminant animal. Basically, if horse is getting into the food chain, no one knows whether it contains bute unless they test. Going further east, of course, the consumption of man's best friend, the dog, is considered perfectly normal. In general, the human consumption of ruminant or non-ruminant quadrupeds is acceptable, but the consumption of carnivores by omnivores such as ourselves is generally considered beyond the pale, especially as it is reportedly done in extremely cruel circumstances.

Horses slaughtered in my locality are treated with great care (I don't know about the rest of Europe or indeed the rest of this country). Each is slaughtered in isolation from others of their species. They are by nature very sensitive animals, and are very aware of strange surroundings and noises. I have seen to my great satisfaction very efficient shooting with a .22 rifle by one particularly skilled slaughterman, who has been killing horses for as far back as I can remember, ranging from huge carthorses to little Shetland-sized ponies.

Apparently, very small carcasses have little value in Europe, and as I understand it most of the small carcasses of moorland-type ponies, which have given many holiday makers and their children such pleasure during the summer months, are rounded up in the autumn and sorted for sale. They fetch very little money at auction and are mostly bought to be slaughtered, and sold to feed the lions and tigers in some of our safari parks, together with the heads of larger horses.

The main horse meat market in this country is catered for by animals such as ex-racehorses, eventing horses and riding horses

– and the bigger the better. Relative to cattle they fetch very little money, and this is of course why the burger scandal has arisen. Quality British horses seem to find a ready market in Europe, and at least these poor devils have not been exported alive. There always used to be a restriction for exporting live horses and it was governed by price: if it was below a certain value it could not be exported. What a good idea! At least it saved them from facing whatever sort of death in Europe.

The next point to consider, particularly with British horses, most of which have led very cosseted lives with no expense spared when it came to veterinary care, is that it is just possible along the way that a treatment of bute may have been administered. Such treatment should be declared and details displayed on the individual horse's passport. Of course, this all depends on the honesty of the owner concerned.

Passports, on which the details and movements of individual cattle are noted, are issued by the British Cattle Movement Service (BCMS). Any bovine animal in the UK may be traced very quickly at any time. Not so with horses: they have 28 different issuing authorities, making it very difficult to trace anything much at all. And those are just the responsibly owned animals; what about all those apparently ownerless animals that we constantly see wandering or tethered on any piece of apparently uncontrolled land? Of course, one cannot say too much about these in case one irritates or offends one of our much loved ethnic groups.

Theoretically, any animal which has been treated with bute is automatically barred from the human food chain, but how do we know? As I have said, virtually all horses slaughtered in this country find their way onto the European market. Inevitably, the best of these animals, as with cattle, will be retained for the steak and quality joint market, but what about the lesser cuts and manufacturing tissue? After what we have heard recently about the shadowy dealings going on throughout Europe and beyond, maybe as far away as South America, it is definitely possible that British horses go to Europe as the first move, but there is no reason why the trimmings don't come back again amongst the vast quantities of horse flesh coming out of particularly Romania. Apparently, Romanians have banned the use of horses to pull carts along the roads, making the horse redundant. This of course has reduced the value of a horse, and if one considers a New Forest pony in the UK can fetch as much as £5 it is evident that the vast numbers of redundant servants have suddenly become food.

When large profits are to be made without many questions asked criminal minds home in on the bounty. I have even heard rumours that the Mafia may well be involved in the game of musical chairs in Europe. Full marks to the Irish watchdogs who were obviously streets ahead of our own Food Standards Agency (previously the Meat Hygiene Service), who have been found to be seriously wanting when the chips were down. Perhaps if their hierarchy spent less time on self congratulation and more time teaching inspectors in the old-school style, not bringing in pernickety form-filling exercises to justify their existence, they might possibly just about be ready for the next fiasco.

Talking to friends since the graphic exposure of the horse meat fraud, some would apparently be perfectly happy to continue to eat whatever the supermarkets and their suppliers are prepared to dish up. Most, however, have a natural revulsion at the idea of eating what are perceived as pets or servants. It has to be said that the consumption of horse flesh will continue in Europe as it always has, so why don't the dealers in this commodity bite the bullet and do what they should have legally done in the first place, which is to clearly mark all those products which contain horse flesh? They will then be able to continue with their cheap and cheerful products and the consumers can make up their own minds.

People do not like to be lied to, but have got used to smoke and mirror tactics from our politicians. They seem to be showing surprise that big business, which all parties cuddle up to just in case there might be a place on their board of directors for a retiring politician, does the same. This latest debacle only reinforces my opinion of them. You only have to watch them squirm out of any sense of responsibility by responding to questions posed by reporters by apparently making it "absolutely clear", "very clear" or even on occasion "crystal clear" or maybe just "clear". As soon as the clear word is mentioned by a politician you can be sure they are trying to muddy the water on whatever topic is under discussion at the time. They must all go to some school which teaches them how to avoid the straight question. They have obviously all been taught by the same teacher because they *all* come out with the same evasive words.

Where in the hell were they and their appointed FSA chiefs whilst all this subterfuge was going on? Not just the horse flesh scandal but the general subterfuge of our reconstituted foods which has been going on for 40 years in my experience alone, ever since the "big boys" took over the meat industry from the small suppliers.

The job I have known for so many, many years has been to protect the public health. Our existence was not known to many of the general public, but I like to think that we have made a difference. That job seems to be largely slipping away in favour of acres of forms and general paperwork often brought into being by people who know little of the intricacies of the meat industry but wield more power than is good for them.

There are still some indigenous properly trained inspectors left, but it seems that today it does not matter to the big veterinary companies how good an inspector you are, what matters is how good a form filler you are, to make as much money for your company as possible. It breaks my heart to see the low level of morale amongst my erstwhile colleagues, and how the job has degenerated and diminished. The main agency involved was so ignorant of what has been going on over the years in the big business side of the meat industry that they failed miserably to test the resultant product fed to millions of people. They are too busy self-congratulating and dreaming up new nothing jobs for the hierarchy to be paid exorbitant amounts of money. This is of

MISREPRESENTATION 153

course not confined to the meat industry; it is in almost every walk of life today.

Another classic hit the headlines again recently, involving the commercially produced curries now available. It would appear that what was described on the tin as lamb curry (probably in fact old ewe and ram curry) in many instances sampled did not contain ovine flesh at all. Would you believe beef and pork were the basis of it, and of course we all know that porcine flesh is a no-no for a large proportion of our citizens.

Organic farming has become very big business over the last fifty years. People like to think that they are eating wholesome food without added poisons, anti-biotics and various other chemicals. Inevitably the cost of producing such unadulterated food is going to be much higher than that which is produced in what has become the generally accepted manner. I am sure that the vast majority of those producers who go down the organic route do so with pride in their products. However, inevitably there will always be someone who is prepared to try and beat the accepted criteria in order to make extra money by being devious.

About twenty years ago, when working at Nailsea, the owner contracted to slaughter organic cattle for the burgeoning organic industry. Various criteria had to be observed on the part of the abattoir. The resultant carcases were then taken to a processing plant in Milton Keynes. The business partnership worked very well for some time, and the organic producers would bring their animals to Nailsea, usually two or three cattle at a time from each producer. Many of these cattle were of the rarer and more original breeds, like Highlanders or Old English Longhorns, which did not necessarily produce the best of conformation when it came to carcase saleability, but I am sure that the flavour and succulence would have been exceptional.

However there was one producer who came from the Lower Somerset/Devon area, who was regularly sending in maybe fifteen or twenty cattle at a time. In order to do this, he would have had to be in a very big way of business which although not impossible was at the very least unusual. Bearing in mind that all the food that those animals would have consumed during their lives should have been only organic hay and concentrate, the amounts required for these numbers of cattle would have been not impossible but difficult to obtain.

Apparently organically produced animals may not be given preventative treatment for such things as worm infestation in order to keep the chemical content of their bodies to a minimum. They can however be treated for symptom if they fall ill for whatever reason. One day, I thought I would have a look at the contents of the reticulums of these cattle. The reticulum is the second of the four stomachs to be found within a ruminant animal. Foreign bodies of various sorts if present are nearly always to be found within the reticulum, particularly if there is a certain amount of weight attached to the foreign body concerned. The drug companies are fully aware of this phenomenon and produce various drug boluses, usually to prevent worm problems, but others are available to be implanted in the animal's stomach with an applicator inserted via the oesophagus. There are also magnetic boluses implanted to hopefully trap ferrous foreign bodies such as nails, staples and pieces of wire etcetera, which may accidentally have been swallowed and present very real risk of penetrating a part of the digestive system, in particular the reticulum, leading to a condition known as traumatic reticulitis which in turn can lead to a horrendous infection throughout the abdominal cavity.

Returning to the drug boluses, they are very cleverly designed with a metal weight at one end, and a number of plastic cups full of drugs attached to it in the form of a tube. The cups are designed to detach at intervals and disperse their drug as a preventative wormer. After a period, all that is left in the reticulum is the weight. Only when the animal is slaughtered does the weight become evident again.

When I took a look at the contents of the reticulums from the cattle from this particular so-called organic producer, I found a number contained these weights. Knowing that this preventative treatment was not allowed for organic cattle production, I wondered why an organic farmer would put his business at risk by using these drug boluses. As this particular aspect was none of my business, being in the cattle killing business and not in the ethics of organic farming. I decided to say nothing about my findings.

However, it is said that your sins will eventually catch you out, and in this instance, the day arrived when a representative from the organic watchdog came to see how well we were dealing with their cattle. He was very good at his job and after watching the Nailsea process he went into the adjacent room, where the green offal i.e. stomachs and intestines are processed. Like me before him, he examined the stomach contents of several of the organic cattle from the big producer and would you believe found as I had several weights from drug boluses which had been implanted some time before.

It appeared that this particular farmer, in order to boost his numbers of "organic" cattle, had been buying in large store cattle from other farmers, finishing them on his own farm and then sending them into us at Nailsea as "organic". Unbeknown to him some of these cattle had already had drug boluses implanted by their previous owners. He had thought that he could boost his numbers and therefore income from the organic premium paid for these non-organic cattle. Inevitably, I heard later that he was struck off the list of organic farmers.

CHAPTER 22

ESCAPEES

OVER THE YEARS there have been many breaks for freedom by condemned animals. One can't help but have a secret admiration for such individuals. You only have to look at the two Tamworth pigs, dubbed the "Tamworth two" by the press, which escaped the abattoir. They were given the names Butch and Sundance after the famous American outlaws. They captured the public imagination to such an extent that funds were raised to keep them alive. Butch died in 2010 aged 13 and Sundance died in 2011 aged 14, having been kept in a rare breeds animal sanctuary in Kent.

Most animals that escape only get out into the slaughtering area. Many are very afraid, and therefore very aggressive, and will charge everything in sight, even the dead bodies of other cattle in the case of bovines. Occasionally, breakouts will occur outside the abattoir curtilage. One cow I remember escaped at Gordon Road, and went into gardens alongside the road, jumping all the garden fences. We gave chase in my Triumph Herald convertible. By the time we caught up, laden with ropes etc to try and capture the escapee, the cow had found its way onto a playing field behind the houses. It was very aggressive, and would even charge the car as we tried to get close enough. It was like a bullfight with passes made by the car as the slaughtermen, who were sitting on the edge of the car with the roof down, tried to lasso the animal. Eventually, one rope and then two were around its neck enabling the slaughtermen to restrain and hogtie the escapee, sadly, not before it had pressed in one of the panels of my car. The animal was taken back to the abattoir to be rested and calmed down prior to its eventual slaughter. It is not good practice to kill a highly stressed animal for human food because the meat can be tough to chew through and can appear blood congested i.e. an off putting red colour.

Another cow once broke out from Gordon Road and ran along an abandoned railway line at the back of the abattoir. It was a bitterly cold frosty morning as I remember. Louis Thorn, one of the meat inspectors, volunteered to accompany a police marksman along the escape route. After some distance they came to an overgrown area on the side of the track. The cow had backed herself into some of this heavy scrub, and because of the cold temperature, all that could be seen were two jets of steam resulting from her exertion running up the track for a mile or so: it looked like something out of a cartoon comic!

The police marksman raised his rifle, preparing to shoot. Louis said "What are you doing?" The police officer replied "I am going to shoot it." Louis then said "Hold on a minute and give me time to return down the track for a few hundred yards." "What's the matter?" said the officer. "You can't even see the beast, and if you wound it, it will come out of there faster than you can run," responded Louis. The officer decided that discretion was the better part of valour and waited until a clear shot became available.

When stressed like this, domestic animals revert very quickly to wild instincts. Wounded buffalo in Africa will do the same thing, to try and ambush their tormentors. I have seen this behaviour several times during my life.

There was a very small slaughterhouse right out on the northern edge of Bristol. The owner saw a niche market with the increasing number of Asian immigrants, who were known to have a penchant for goat meat. In order to satisfy this demand this man started to import goats from Southern Ireland, where apparently they run wild in certain areas. Allegedly, the method of catching these semi-wild animals was to put a large net at one end of a village street, block off all the side alleyways, and drive a whole flock of goats down off the hills into the street, closing off any retreat with another large net across the other end of the street. The animals were then chased around until all were caught. The consignments contained all ages of goats, from little kids to huge Billy goats with very large horns; in fact, when packed together most of what you could see was a forest of horns. I still have a pair of these huge horns today!

The lairage at this tiny abattoir was composed of little more than converted pig sties with netting over the top. Inevitably, at various times goats would escape these confines: they were very agile creatures, and able to jump like deer. A small flock used to roam the locality, decimating local allotments and gardens. It got to be such a problem that a marksman had to be called in to dispatch them; this was accomplished over a period of time because one shot would spook all the others.

On another occasion, just before Christmas, a steer with fairly long horns escaped from Gordon Road and ran down a main street crowded with shoppers. Unaware of the danger, the shoppers were pointing out this "cow" to their young children. I could see a tragedy coming. Again, four or five slaughtermen sitting around my Triumph Herald clutching some hastily gathered up ropes set off in pursuit. We soon found our progress barred by a traffic jam. Knowing the reason for this jam I pulled over to the other side of this main road to overtake all the cars. There were some "keep left" bollards in the middle of the road which I ignored in order to catch up with the steer as quickly as possible. Inevitably, a siren and blue flashing lights were soon seen and heard behind me, and I was flagged down by a quite belligerent officer of the law. "We are the only ones allowed to do that," he said, to which I replied "Well in that case you had better be the only ones to catch this steer before somebody gets killed." He changed his attitude at this and escorted us on down the road, blue lights and siren going.

When we arrived on the scene it was reminiscent of when Spanish bulls are run through the streets of Spain, prior to a bullfight. People were running everywhere, the steer first chasing one and then another. Miraculously, nobody had been hurt. We managed to divert the steer from the main road onto a small housing estate where, with the aid of a van, a police car and my car in the rear, we corralled the beast against a wall. It decided to try and make another break for it, by trying to climb out across the bonnet of the police car. Peter Worle, then a very strong young man who used to engage in international tug-of-war tournaments, grabbed the steer by one and a half horns (it had broken off half of one horn charging a car) and threw it to the ground in true rodeo style, whereupon we all jumped on it and trussed it up like a chicken so it could not move. We then waited for a cattle lorry to collect it and return it to the abattoir.

While waiting for the arrival of the lorry a police marksman appeared and, with total disregard for the assembled onlookers who had collected in droves, raised his rifle and blew a large hole in the steer's head. I confronted him and asked him what he thought he was doing. He said "It was dangerous, and I had to shoot it." "It could not move, being trussed up as it was; you could easily have had a piece of bone or tarmac ricochet and blind one of the onlookers" I replied. "I was using special bullets," retorted the marksman. Being a shooting man myself I was fully aware of the dangers posed by soft-nosed bullets. This officer had been issued with a firearm, and was intent on using it with total disregard of the dangers.

There have been several occasions when the Police appear to have been too quick to use their firearms. A little bit of power in the wrong hands is a very dangerous thing.

Although the animal had bled a lot through the hole in its head, if it was going to be used at all for meat it had to be stuck. In the excitement of the chase none of us had brought a knife with us. I asked a woman in one of the houses if I could borrow a carving knife. She produced a knife, sadly with no edge on it. I had to saw my way through the steer's throat. After the woman had witnessed what I had done with her carving knife she did not want it back, so I wiped it on her lawn and left it on her window sill. We dragged the steer into the back of a cattle lorry and returned to the abattoir. The resultant beef would have been of very low quality because the animal had been so stressed prior to death, and would probably cut very dark. Its eventual use would probably have been for manufacture.

While working at Weston-super-Mare abattoir a beast escaped and lived wild for about a fortnight in the woods surrounding the town, coming out at night to run along the beach. It was eventually caught and, again, money was raised to buy it and keep it alive.

At Nailsea there have been a few escapers in the slaughterhouse. In order to shoot them with a captive bolt pistol one has to be right up close to the animal's head. If the animal is in a fractious state it is just not worth the risk to the slaughterman. The owner at Nailsea knows that I have both a rifle and pistol licence for food animal slaughter. If he gets an escapee, he will give me a ring. By climbing onto the high level platforms I can be safe to shoot down onto the animal underneath, even if it is trying to get at me from below.

There was one escaper from certain death at Gordon Road in Bristol. This was a large ram which we called "Teddy"; this in fond memory of Tom (Teddy) Howick. There was a contract in the later years of Gordon Road which, if it could be filled, was for 1,000 rams per week. These, together with around 2,000 ewes, were destined for the curry trade, because breeding rams tend not to carry much fat, but are full of muscle, albeit very tough. The trade did not want any fat at all and so these big rams fitted the bill very well.

Emaciated badger with TB.

Badger head with tubercular abscess.

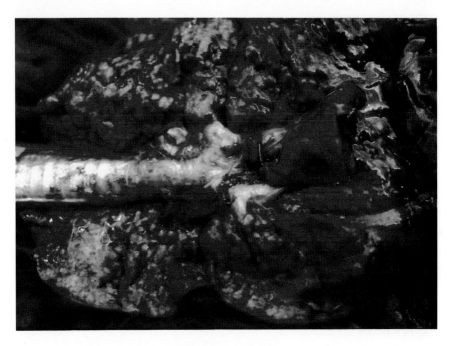

Badger lungs with TB.

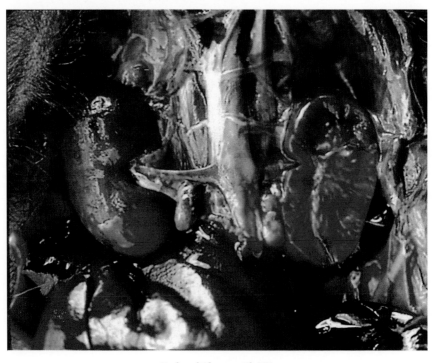

Badger kidneys with TB.

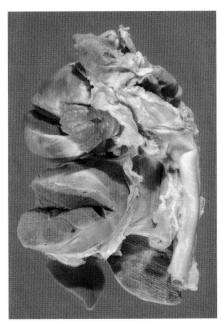

Pig with two hearts.

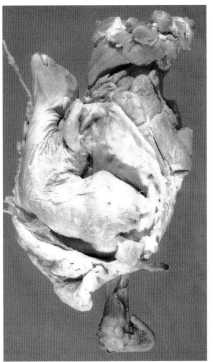

Lamb leg with complete foetus developed inside it.

Conjoined calves

Examples of the terrible condition that some old sheep were in when arriving at Gordon Road destined for the "curry trade".

Fallow deer, which had been trying to walk on two fractured front legs. Maybe as a result of a RTA (road traffic accident) or maybe getting its legs caught in between two strands of a barbed wire fence.

Roe deer which had been attacked by someone's pet dog and bitten all over. Many people do not realise that they have a serious carnivore under their control or, in many cases, not. This animal was still alive and had to be shot.

Four lovely carcases of beef which were rejected for human consumption and slashed and stained at Nailsea.

Early days at Gordon Road.

Training at Langford.

Teddy Howick who amongst others taught me so much.

Teddy Howick again.

Oestrus Ovis (maggots) within the sinuses of a sheep.

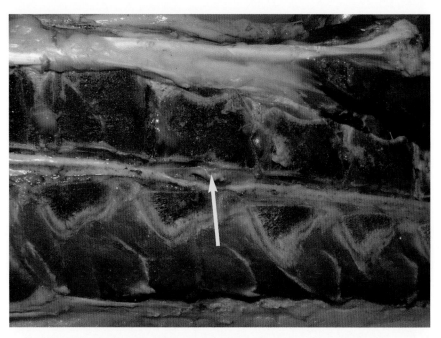

Pig spine with a small abscess where infection had spread from a bitten tail.

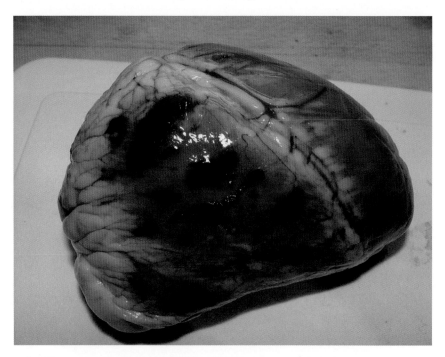

Sheep heart with blood splashing due to too high a current during electrocution, or too strong a charge when shooting and too long a delay between stunning and sticking.

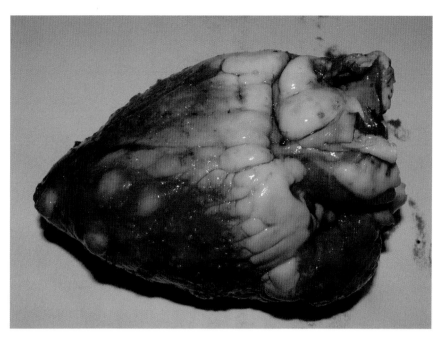

Sheep heart with multiple cysticercus ovis cysts.

Cattle lungs with melanosis, a depositing of the body's black pigment where it should not be.

Sheep intestine with tapeworms coming out.

Sheep liver with liver fluke.

Sheep lungs with lungworm.

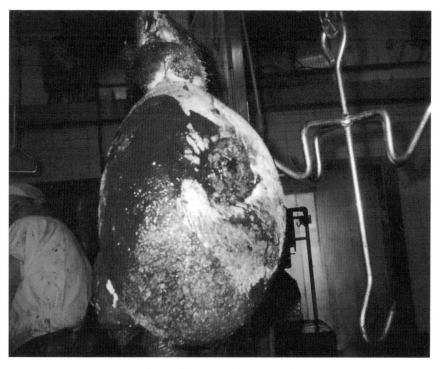

Huge ulcerated abscess on lower jaw of a cow.

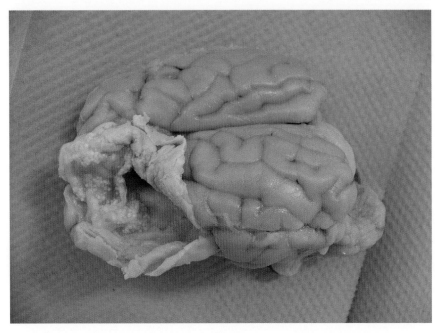

Sheep brain with a multiceps cyst, a condition known as "gid" or "sturdy", yet another intermediate stage of a dog tapeworm, each cyst contains hundreds of potential tapeworm heads "ready to infect a dog when it consumes the brain".

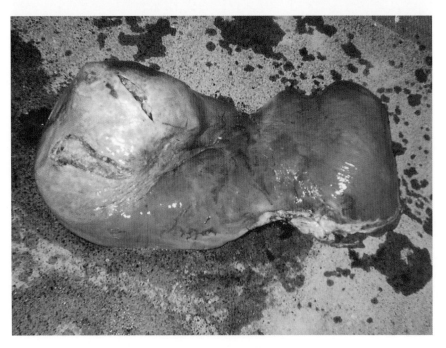

Huge abscess in a cattle liver.

A birth deformity of the front legs, this animal has not thrived due to its condition.

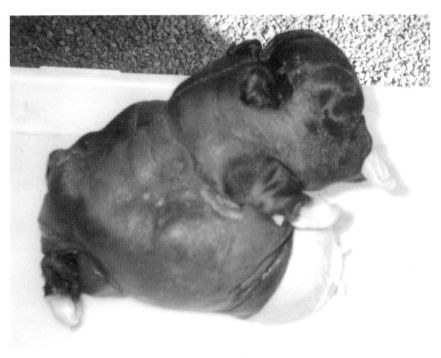

Unborn "bulldog" calf found within the uterus of a Dexter cow. It has an undershot jaw, flippers for legs and some internal organs outside the body.

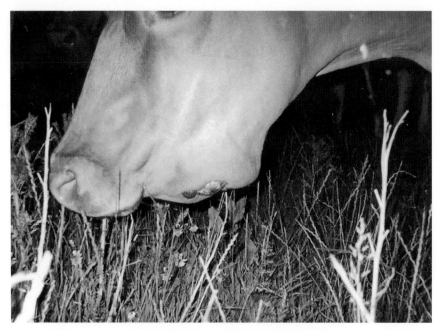

Guernsey cow with a "lumpy jaw" actinomycosis which destroys usually the lower jaw but occasionally the top jaw.

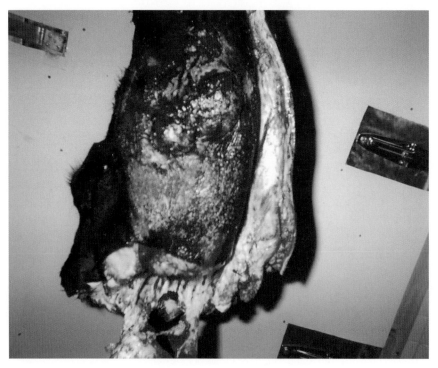

Another example of actinomycosis.

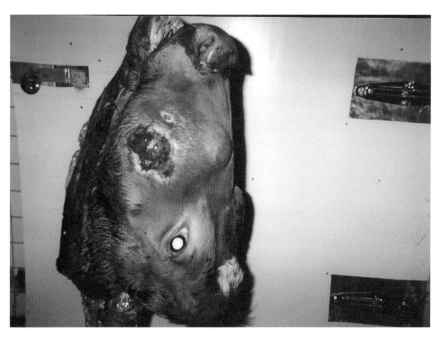

Actinomycosis can occasionally be in the skull itself.

Large bull slaughtered during the BSE cull. They were sometimes so large they would not fit into the stunning trap so they were shot outside and dragged in.

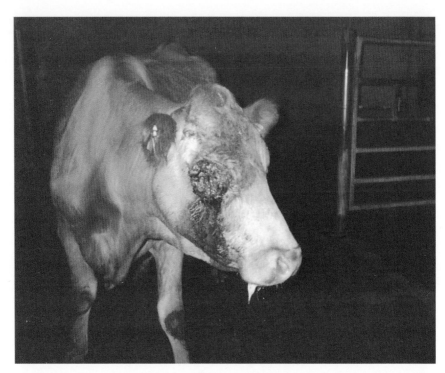

Cow with a tumour in its eye.

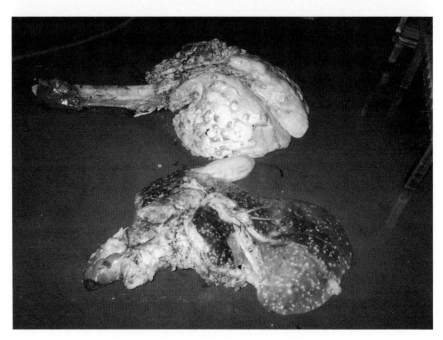

Cattle liver and lungs with multiple TB lesions.

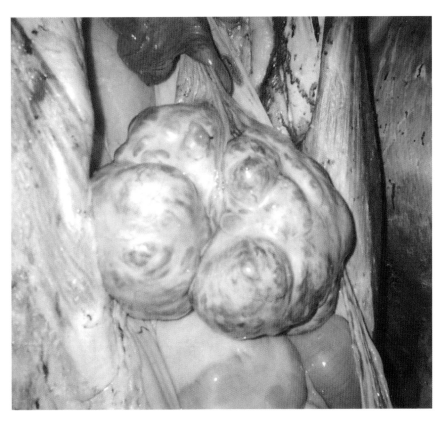

Huge tumour in the ovary of a cow.

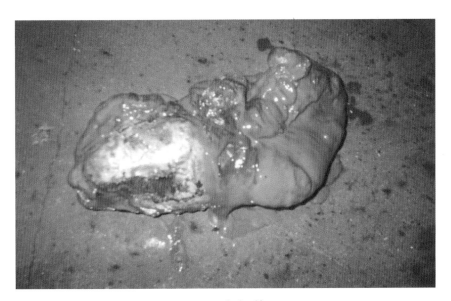

Mummified calf.

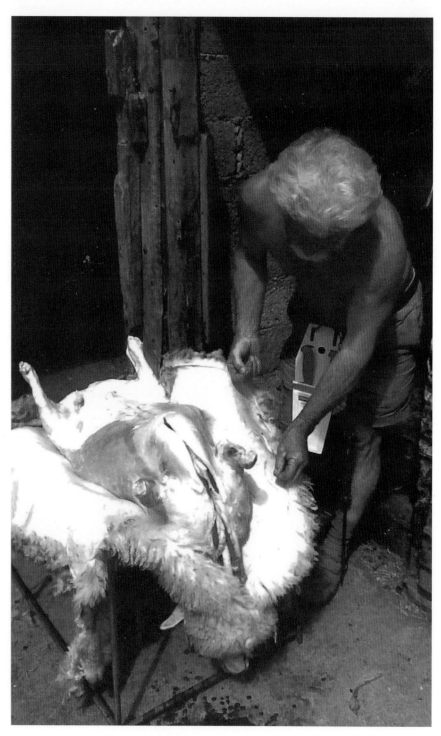
Home dressing for the freezer.

Home teaching in front of my dog kennels.

Preparing for a student exam in Cardiff.

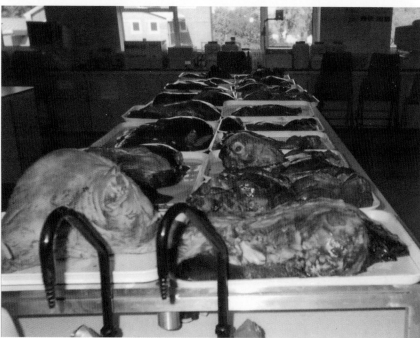

Examination tables of meat and fish laid out in Cardiff where I used to examine environmental health students, as I did at Langford.

Old fashioned equipment.

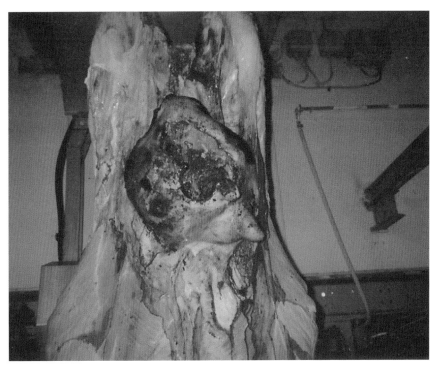

Mastitis in a cow's udder.

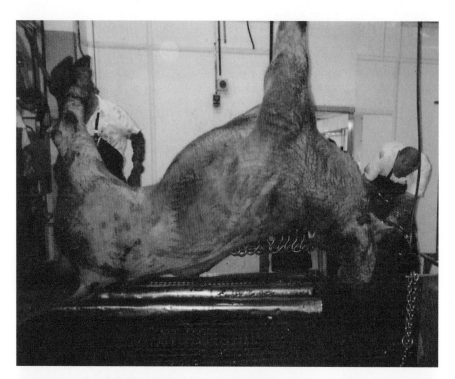
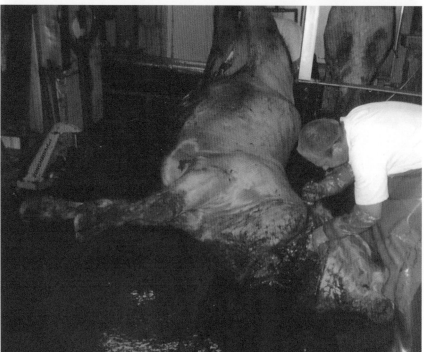

The workplace at Bridgwater, dressing a bull.

Tim and Steve inspector friends.

Environmental health students with me at Nailsea.

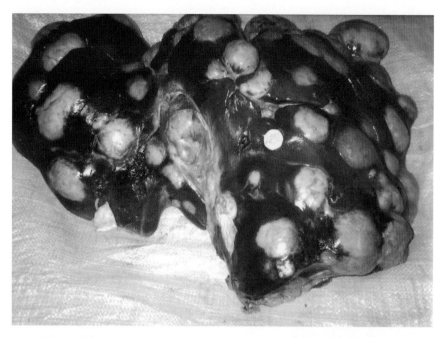

Catlle liver with hydatid cysts, another dog tapeworm in the intermediate stage. Dogs infected with this tapeworm can infect us at this stage. **Never** let a dog lick your face.

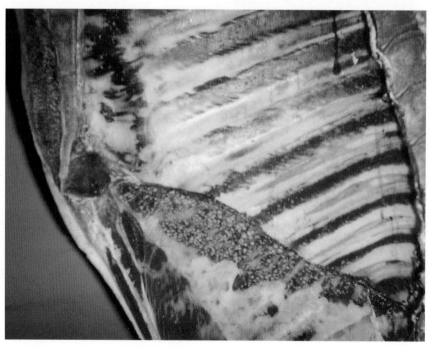

Beef carcase with mould growth, this is due to temperature fluctuation in the fridge.

Cattle flesh with a viable cysticercus bovis cyst. The cyst has been removed from the capsule and the head is the white spot inside the cyst bladder. If consumed by a human this can develop into a twenty odd foot long tapeworm called taenia saginata.

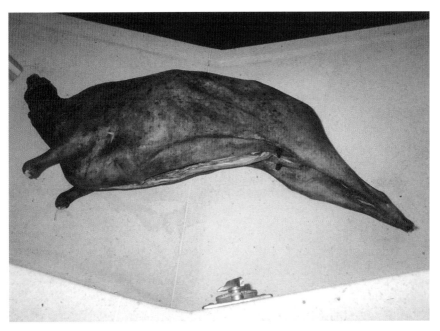

Carcase of a sheep dressed illegally by having the wool burnt off for the particular consumption of ethnic groups. Known in the trade as a "smokie".

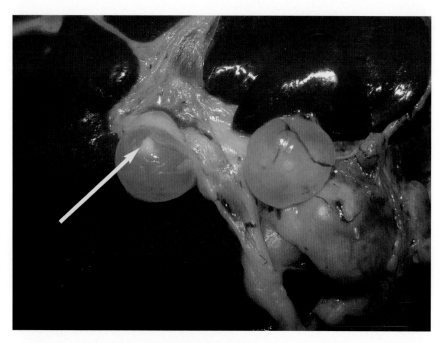

Sheep liver with cysticercus tenuicollis cysts. This is the intermediate stage of the dog tapeworm taenia hydatidgena. The potential tapeworm heads can be seen as white spots within the cysts.

Pigs liver with damage from the migration of larvae of the intestinal roundworm of the pig called ascaris suum. This condition is known in the trade as "milkspot".

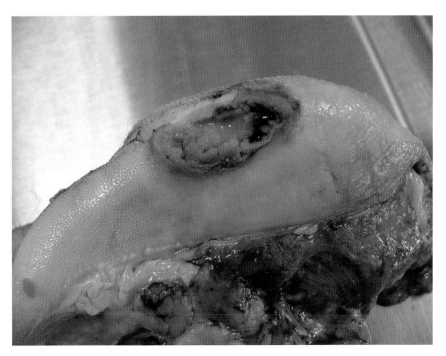

Calf tongue with ulceration due to calf diptheria.

Animal with broken leg at Nailsea.

Arthur Williams with the very first chine saw, developed from an old motorbike engine belonging to his son, Fred.

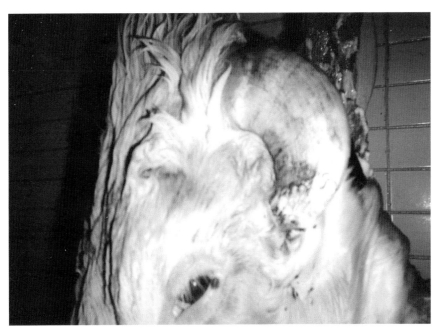

Ingrowing horn causing much distress to the animal concerned. All of which could be avoided with one minute spent with a hacksaw.

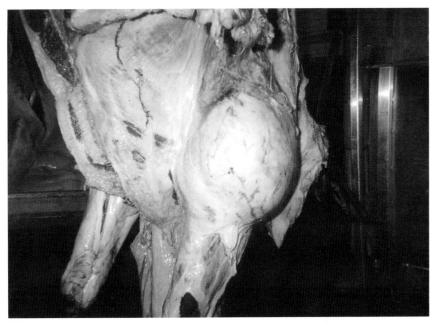

Abscess in beef shoulder.

My natural habitat on the Quantock Hills in Somerset.

I often wondered how these old rams could be eaten. Just pulling their skins off their backs was a struggle for Frankie Davis, who was usually on the end of the moving "legging" cradle; he would hold the skin and brace himself against the cradle as the carcass was hoisted on an electric hoist and let the hoist tear the skin from the animal using himself as an anchor.

When these large numbers of sheep had to be brought up to the killing point from the lairage along a fairly narrow alleyway, very often if the leaders stopped the whole procession of maybe 300 sheep would also stop. Someone had then to go from the front of the flock to move them on again, which only had the effect of unnecessary stress to the sheep. One day, when doing an ante-mortem inspection of the animals in the lairage, I noticed a large tame ram amongst other rams. He came over to talk to me making the little muttering noise that rams make when they are chatting up the ewes. I had the idea that this tame (and at the time friendly) ram could become a "Judas" ram. Animals like this are used across the world to help move large flocks of sheep. I asked the wholesale butcher who owned him what he thought of the idea, and he soon saw how this ram could save a lot of time and effort. Teddy joined the team and seemed to need no training in what was expected of him. He trotted willingly along the large lairage with streams of ewes and other rams coming along behind. He would walk right up into the stunning pen and wait until all the doors, both of the stunning pen and the pre-slaughter pens off the main alleyway, had been closed to contain all the sheep that he had betrayed, and then trot back into the main lairage to await the next batch.

Teddy was fitted out with a collar and chain and a screw spike to go into the ground. When he was not working he would be tethered on one of the grassy banks around the abattoir. On bank holiday weekends I used to take him home in the boot of my car to graze in a paddock on my parents' property.

Very sadly some of the young men working at the abattoir at that time used to amuse themselves during their lunch hours by goading

Teddy into charging them. Hitherto he had been quite a placid creature, but soon he became quite aggressive. Bottle hand-reared sheep like him had no fear of humans, and if they become aggressive it is very difficult to contain them. His owner took him out to his own farm when I was on holiday one year to run with the hundreds of sheep he used to keep there as a reservoir to be drawn on as and when needed. There were some public footpaths across this land and the owner was concerned that Teddy would attack some unsuspecting walker. His solution was to have Teddy slaughtered while I was on holiday. It has to be said that I was not very happy on my return, but in retrospect if Teddy had injured someone the owner did not want to be sued for damages. Teddy was such a character, and it was such a shame that he had to die all because some young men without realising what the repercussions might be had amused themselves by annoying him. Although Teddy only prolonged his life by a couple of years, it was two years that he would not otherwise have enjoyed.

Some years later I owned a Wiltshire Horn ram of my own which, although not bottle reared, became very tame; in the case of rams this can be a big problem. "Rambo", as we inevitably called him, became very aggressive, and if he could score a hit he would almost chuckle with delight. One day he crept up on me when I was concentrating on something else, and in a crouching position. He broke two ribs, putting me through a glass door. It was during the summertime and I was only wearing shorts with no shirt. I managed to fight Rambo off with my feet and made my escape. As I returned to the house I could feel a warm trickle of blood running down my back as a result of some glass shards stuck in my back, necessitating a visit to the casualty department, where I had to explain that I had been savaged by a sheep! Friends at work suggested that perhaps it was time to remove Rambo's head from his body, but I had become quite fond of him and, like Teddy, he was quite a character. He eventually lived to the ripe old age of 16. In the end I had to put him down when he developed a kidney infection, but I still have his skull and horns to remind me of that ill-tempered old character.

Over the years nothing has changed in the way the police overreact when dealing with escaped animals, in particular escaped farm animals. I suppose they are never trained in farm animal handling, and don't often get the opportunity to use the weapons that they have trained with on inanimate targets on living beings, so when an animal escapes from a market or an abattoir, backed up by health and safety legislation, they can shoot with impunity.

In October 2012 I read of an incident in Welshpool, when a very stressed young Friesian heifer had made a bid for freedom from the local livestock market. Police were summoned to the incident, and complete with body armour and safety helmets, and sporting submachine guns, you would think they were about to take on a band of desperate armed robbers, or possibly some fanatical terrorists, rather than a terrified young heifer cornered in a back garden. The decision was taken that the animal was a danger to the public and as such must be destroyed. Apparently, some police "marksman" who trains regularly on a shooting range took four shots to kill it. I have also heard of it taking four shots to kill a maddened pit bull terrier in another area and under different circumstances. God help the badgers if that is the level of competence considered by the police to be termed a marksman. Having personally killed thousands of animals during the foot and mouth debacle, I know if it had taken us four shots to kill an animal we would very soon have been thrown off the job, and rightly so.

CHAPTER 23

BUREAUCRACY GONE MAD

IT REALLY BEGAN just before local government relinquished the reins of meat inspection. We first had the imposition on the meat trade of Official Veterinary Surgeons. It was a heaven sent opportunity for some veterinary practices to make an awful lot of money very easily. In the beginning most of the vets involved were doing the job as well as running their own practices, or several that I knew were retired and earning extra money to embellish their pensions.

Later, when the specialist money grabbing veterinary practices took over, nearly all the vets operating in the abattoirs were young, freshly qualified Europeans. These young people many of whom were out of their depth, were not earning much money and were terrified of the companies employing them., because they were very expendable. There were many others out in Europe waiting for the opportunity to come to this country.

When they are brought into this country by the various recruiting companies they usually operate as meat inspectors to start with. Their veterinary qualification in Europe allows them to do this. Having been asked to try and train some of these usually very nice young people in a very short time, I am fully aware of their lack of knowledge of food animal pathology and its relationship to public health.

Europe were entirely responsible for this, because in Europe they had a different system of inspection, allegedly everything was veterinary inspected. They had no highly experienced, dedicated 'meat inspectors'. Years ago Louis Thorn, one of my lads in Bristol, went on holiday to Denmark. Some people when on holiday abroad like to see how other people do things. Louis went to visit a huge abattoir, killing cattle sheep and pigs on three separate lines, all operating simultaneously.

He was introduced to the vet in charge of this huge complex. The vet was very keen to show Louis around. His office had a large panoramic window overlooking the abattoir floor. Louis asked the vet if everything was veterinary inspected? "Yes of course" was the reply. "Who then" said Louis, "are all those people in white overalls?" "They are the spotters" said the vet. "Who are the spotters?" said Louis. "Well "said the vet, "They are usually operatives made up from the slaughter line" "What is their job?" said Louis. "They check everything for disease problems and if they find anything abnormal they call the vet to make a decision". "Well" said Louis "in that case the job is only as good as the unqualified spotter." The vet looked bemused by this observation.

We have almost arrived at this nonsensical situation in this country now. The most inexperienced young European vets are now doing the job with the comparatively few properly trained indigenous meat inspectors left. (There are currently no courses for meat inspectors run in this country). Some of the horrendous stories I hear from some meat inspector friends and what I have seen with my own eyes make me cringe. The job overall has deteriorated beyond my recognition.

My job has always been about dedication and pride in doing something for the public good. The job is done behind closed doors and there is little awareness among the wider population of our existence. I have found this to my satisfaction when giving talks on the subject to various groups such as Rotary, Round table and various specialist groups who are always looking for somebody different to talk to them at their quarterly or monthly meetings.

The Food Standards Agency (another Government impersonal top heavy nonsense) which absorbed the earlier Meat Hygiene Service, who in turn took over from the good days of local government retain a skeleton staff level and subcontract out to these veterinary practice based recruitment agencies.

Today it is a money orientated exercise and the quality and dedication of the meat inspector is not anything like so important. All that seems to matter now is have you got a H.A.C.C.P. (Hazards Analysis Critical Control Point) system in place and have you filled out the necessary forms?

I remember when I was involved at Langford Veterinary College about 20 years ago, with a group of meat inspector students from the South West who had been doing a course at Salford. We felt that they might need a bit of extra tuition before their final exam, so several of us got together to help them. We had an inspector who was very good at legislation, another who was a poultry specialist and I was doing red meat.

During the course on the day we had a visitation from the head of the Meat Hygiene Service. He was obviously looking at the possibilities of what we were doing, with a view to maybe running a meat inspectors course at Langford and he spent time with all three of us. When he came into the post mortem room where I was teaching, he asked politely if I minded him listening to my talk. I told him to pull up a chair, because he might learn something. I have little time for administrators who have little if any knowledge about what it is that they are dealing with. The man however pleasant had been recruited from the waste disposal industry in Ireland.

He listened to my lecture, and asked to remain for a further lecture with a second group of students. At the end, he came to me and said he had never seen anything like the specimens which I had been talking about in his life. My rejoinder was that he should get out more.

He was quite a nice man, and I could have told him what the job was all about given time. However he went back to his administration and I returned to my work. I think most of us today come up against this nonsensical bureaucracy. You only have to look at the army of administrators that the politicians inflicted on the National Health Service, instead of leaving the doctors and nurses, particularly the old fashioned matrons to do what they used to do so well. These bloody freeloaders earn fantastic salaries but we cannot give our nurses a decent pay rate.

As a result of this high ranking visitation and having seen what was involved in the real job, Langford was asked to provide a meat inspection course, the first of two. Many of those students who were properly taught in the old fashioned way went on to high ranking positions in the Meat Hygiene Service and beyond. At this time the M.H.S became strapped for cash with monies squandered on useless morale boosting weekends, using meaningless words like team building. Large salaries for pen pushers, too many area offices (now closed) and millions of pieces of paper.

In the end they decided they could no longer afford to pay for the best meat inspection courses run in the country and they settled for cheaper and inferior courses. The final one was run at Harper Adams Agricultural College in Shropshire. I was asked to take some morbid specimens and travel there to do some lectures. When I arrived I ran into an obnoxious jobsworth, who wanted to see movement licences for unfit meat. He was unimpressed when I told him I had none of these, and that I would be returning all the fresh material from whence it came. He said that he must have the necessary paperwork for his control system.

I told him that if he persisted with his stubborn bureaucratic stance I would turn my car around and return to Bristol and he could explain to the students and those running the course what had happened. He considered this for a minute, and decided to back down.

The fact is that I take similar material to Langford Veterinary College on a regular basis for student training, after which it is incinerated in their private incinerator. It is under my control or the control of the staff of the pathology department all the time. These people, who are apparently born without a grain of common sense, are just what the bureaucrats love, to enforce their crazy dictats. I despise every one of them.

The students welcomed me with open arms. They told me that during the whole course they had seen nothing like that which I was showing them. This tells me that this course amongst others was not covering this very extensive subject of meat inspection thoroughly. This is an unacceptable situation for potential meat inspectors from this country and I experience a similar situation when I encounter the young European vets from universities abroad who I have to train to undertake the work of a meat inspector – a position that they are meant to supervise.

The European vets are given a flat, given a car, given a knife and told to get on with the job with little or no experience and in a foreign country from that which they are used to. The deterioration of the job is definitely not their fault; with careful tuition they would probably make good inspectors. But this is the way the job has been allowed to go with the introduction of new European legislation which insisted on so called veterinary inspection in abattoirs. The fact that we in this country had a perfectly efficient system run by meat inspectors was of no interest to Europe. I fear that our system as with so many good things in this country has gone forever.

In my young day the old inspectors (a bit younger than I am now) would call the young inspectors over to look at this or that condition, which would come along occasionally and each would be discussed in detail. Gradually one would gain confidence and experience. What better way could there be to learn about what is a very wide subject than this? I feel very sorry for these young vets today; they are not even taught how to sharpen a knife.

When working at Nailsea be it for the local authority or for the M.H.S. it was easily within my capabilities to cover all aspects of the job. All that has changed is bureaucratic nonsense and paperwork but it seems it now requires two inspectors and a vet to cover a lesser kill than in my day. The question to be asked is, has the product improved? I think not. Has the co-operation of the abattoir management to the bureaucrats improved? I definitely think not. Has it created lots of pen pushing jobs? Certainly. The additional costs to the industry must be astronomical.

CHAPTER 24

RITUAL SLAUGHTER

TOWARDS THE END of the 1960s and throughout the 70s the Muslim population in this country was increasing rapidly, and the demand for Halal ritually slaughtered meat was increasing along with it. The owner of the abattoir at Gordon Road took the decision for the abattoir to become one of only a few plants in the country at that time to cater for this demand. From a business point of view this was a very shrewd move, as to the best of my knowledge at that time there were no Muslim-owned or Muslim-run plants. Today, this aspect of the meat trade has become a huge business, predominately owned and run by the Muslim faith.

Halal slaughter had to be performed by a Muslim, so the decision was taken to employ a man by the name of Ali Iqbal who, between batches of sheep, did some abattoir cleaning to ensure his working day was filled constructively. It was agreed very early on with the wholesale Muslim butchers that all the sheep would be electrically stunned, and the only difference between Halal and traditional slaughter was that the actual throat cutting was performed by a Muslim and accompanied by a prayer. Theoretically, the animal should be able to hear this prayer while its throat is being cut. Electrical stunning as used at Gordon Road appeared to be acceptable to the vast majority of Muslim customers, whereas the shooting of sheep with a captive bolt pistol was definitely unacceptable.

However, after some years of this arrangement electrical stunning was also called into question by some of the mosques in London. A request came through to me as the senior inspector for some of the high ranking members of the mosques to witness the electrical stunning of a sheep, and to watch the animal recover and rise again to prove to them that it was not dead before throat-cutting. I held a discussion with the other inspectors and then decided that the gross discomfort of one sheep would be to the benefit of many thousands that came after, if it was decided that electrical stunning was still acceptable.

On the appointed day six or seven representatives duly arrived and one unfortunate sheep was driven to the stunning pen and, with the representatives on one side of the pen and the inspectors on the other, duly stunned. It lay on the floor with its feet drawn tight to its body in muscle spasms which, after a few moments, could be seen to start to relax. Gradually the sheep tried to rise to its feet once again, but was falling about the pen as if drunk, whereupon it was stunned again and slaughtered in the normal way. This demonstration, however distressing, achieved the desired result, and the concerns that had been raised were now laid to rest.

This demonstration was illegal, but I stand by my decision. I have spent my working life trying to stop people from being cruel to food animals; in fact, the only prosecutions I have ever taken have been on animal welfare grounds.

Not all the throat-cutting was done by Ali Iqbal, occasionally one of the Muslim wholesale butchers wanted to kill their own sheep. On one such occasion I was met by a Muslim butcher at half past six in the morning on my arrival at the abattoir. The instruction was, if such an occasion arose, that any such person who had been issued with a slaughtering licence elsewhere in the country was to see me before he started killing, first to see that everything was as it should be with the licence and secondly for me to judge his capability at the job.

This particular man was a fairly aggressive and unpleasant individual, and he resentfully presented his slaughtering licence for my perusal. It had been issued by one of the London boroughs. Everything seemed to be in order: the licence had been granted for Halal slaughter with the use of a knife. This individual proceeded to tell me that he wanted the sheep he was buying, some 300 a week, to be hoisted by a back leg and he would then cut their throats in the air. I looked at him as though he had just crawled out from under a stone and told him that no way in the world would unstunned sheep be hoisted alive for him to cut their throats. I told him that because of his licence I was not able to stop him killing the sheep, but we would proceed with the slaughter by having one sheep at a time being placed on a "legging" cradle and held by one of our slaughtermen, while he cut the throat and said his prayer while facing in the right direction. "Then the stunning pen will be cleaned down to my satisfaction," I said. "What do you mean?" responded the Muslim butcher. I said that when every last vestige of blood had been washed away he could kill another one. "It will take all day to do 50," he said. I said "Will it?" He turned on his heel and climbed back into his lorry, gave me the expected two finger salute, and drove off, never to be seen again. Roger Hendy never reproached me for my stand against abject cruelty in the name of what was considered at the time as an alien religion. Today we lean over backwards so as not to offend the sensitivities of these cultures and religions.

I had won these small battles in the persistent war, but I knew that wherever this individual had been killing his meat before he was probably going to return, either to that abattoir or another elsewhere where the senior inspector obviously did not have sufficient backbone to stand up to him on a matter so basic as cruelty to dumb animals.

I have never witnessed, other than on a television programme, the ritual slaughter performed by the Jewish faith. "Schechita" is the name of this procedure, which is performed by a specially trained "shochet" and the square-ended knives used are kept exquisitely sharp. The shochet is assisted by a man called a "sealer" or "shomer", who is responsible for putting the "kosher" seal on each part of the carcass which is considered fit for Jewish consumption, not dissimilar to the health mark which meat inspectors place upon carcasses killed in the normal way.

The use of a casting crate was in order to present the animal's neck to best advantage when cutting its throat. The animal would be driven into the metal crate, which could then be completely rotated so that the animal was in effect lying on its back with the head protruding and locked in a "V" shaped restraint. Later it was considered to be cruel and stressful to restrain an animal on its back. Another form of crate came into use into which an animal was driven. Its head and neck protruded from the crate which could be moved forward on rails and the head and neck introduced into a separate room, where awaited the shochet ready to wash the area of the neck which was to be cut. I have heard that the slaughtermen who loaded the animal into this crate in one abattoir at least disliked the whole procedure, which if they had done it themselves would incur a heavy fine or possibly a prison sentence, they not having the religious derogation necessary. They would withdraw the crate from the cutting room and shoot the unfortunate animal as soon as they could after it had been cut to reduce the suffering in its last living moments.

However carefully the procedure is undertaken it has been known for the unfortunate animal to remain conscious for some minutes after being cut. The reason for this can be due to the elasticity of the carotid arteries in the neck, which when severed have the ability to seal themselves and maintain blood in the brain, allowing the animal to remain conscious for far longer than it humanely should.

There are five principles of schechita. In their traditional order they are that the neck incision shall be completed without pause, pressure, stabbing, slanting or tearing, while the incredibly sharp knife used must not be damaged in any way during the process. If any of the above occurs the animal concerned is not permitted to be used for Jewish food. Any animal which will not rise to its feet when "struck" with a stick may not be slaughtered for Jewish food; likewise any animal which does not move during the slaughter process.

This is of course the first known acknowledgement of meat inspection and the awareness that food animals can suffer from diseases and conditions which can harm the human animal when consuming the flesh. They did not stop at making sure the animal concerned was not moribund, ie dead on arrival. The dressing process would then proceed until it becomes possible to cut around the diaphragm and expose the thoracic cavity, to ascertain as to whether there are any adhesions between the lungs and the thoracic wall (pleurisy). Probably the origins of this practice were to find evidence of bovine TB. Great care is taken over the whole procedure, and afterwards the removal of the major blood vessels (a process known as "porging") takes place. Apparently it is easier to do this in the forequarter than in the hind quarter, which is not often undertaken.

The porging of hind quarters can only be done by very skilled kosher butchers primarily because there are 50 major blood vessels which must be removed to make the hind quarter kosher. I remember ritually killed but not porged hind quarters were sold in the meat depots in Old Market Street, Bristol. They came, I was told at the time, from a Cardiff abattoir where ritual Jewish slaughter was undertaken. These hind quarters were sold to be consumed by "gentiles" and I often wondered if the people eating them would have been so complacent if they had known how painfully those animals had died. However, it must be said that ritual slaughter, when done by the Jewish method taking great care with regard to the sharpness of the knife used, is in my opinion preferable to the Muslim Halal slaughter when, as I have witnessed with my own eyes, lowly skilled people, with any piece of hoop iron in the way of a knife can saw their way perfectly legally through an animal's neck.

The probable reason behind the rejection of pig flesh for human consumption by both the Muslim and Jewish faiths is the presence of the intermediate stage of a human tapeworm called taenia solium. This problem is a common occurrence where human sanitation is poor and the segments of the tapeworm in the human gut, which are full of eggs, are passed out in human faeces. The faeces can in turn be consumed by roaming pigs and the parasite encysts itself within the pig's musculature, waiting to be consumed by humans. If the pork is cooked insufficiently well the life cycle of the tapeworm is completed. How the wise old men years ago made the connection that eating pork might give you a tapeworm we shall never know.

The law in this once lovely country of ours quite clearly states that a food animal or bird shall be rendered insensible to pain at the point of slaughter, however this is achieved, be it by the use of electricity, by gas, by brain penetration with a captive bolt pistol or occasionally by shooting into the brain with a free bullet, the latter being my own personal preference but sadly not available to everyone. In fact if I become stricken with some unpleasant disease or condition in my later years this will be <u>my</u> escape route, having witnessed over so many years the instant relief from pain and suffering. The latter is allowed by law to continue with so many of our fellow humans; when the final end is inevitable why make so many suffer pain and distress? You would probably be prosecuted for cruelty if you allowed man's best friend to suffer in this way and have you ever noticed what his name spells when spelt backwards? I take it upon myself to do the kind thing for my dogs when life eventually becomes unpleasant, my thinking being that I have hopefully given them a good life and then a good death when the time comes. It would be good to know that someone would do the same for me, but I will probably have to do it for myself, because the laws dreamed up by our politicians will not allow any of my good close friends to do this kindness for me, however capable they might be, for fear of prosecution for murder.

Food animal welfare concerns in this country clearly state that to cause an animal avoidable excitement or pain is an offence and research indicates to stun an animal prior to slaughter is far more humane, than not to do so. Although the Farm Animal Welfare Council (FAWC) which is the advisory body to the government of the day has actually recommended that the whole sorry business of ritual slaughter is banned in this country as it is in Norway, Switzerland, Sweden and Latvia, sadly no action has been taken. A bill was proposed in the Netherlands to ban this cruel practice but was defeated in their Upper House in June 2012 due to pressure being brought by Jewish and Muslim groups. Needless to say our own pathetic politicians did not heed the advice given by the FAWC or indeed as if it were necessary, the additional backing of the RSPCA, Compassion in World Farming, the British Veterinary Association and the Federation of Veterinarians of Europe, to name but a few. Apparently the present government does not keep records of religious slaughter and stated in October 2012 that it did not know how many Halal abattoirs there were; as usual obviously terrified of religious discrimination they ducked the issue.

In May 2012 the Food Standards Agency, who are normally obsessive about figures and paperwork, published the results of a survey set up to find out numerically how many food animals slaughtered officially were not stunned prior to slaughter. The result of the survey was that 3% of cattle 10% of sheep and goats, and 4% of poultry were slaughtered without stunning. Those were the official figures which did not allow for the unofficial figures, and as I have seen during my life's work in the St Paul's district of Bristol, goats have been bought in markets and elsewhere, then having their throats cut in back yards. Poultry bought at poultry and bird auctions are taken out to the back of the auction market to have their throats cut before being taken home. But, does it really matter what percent are treated in this barbaric way, surely one is too many?

The industry itself has noticed how large is now the demand for both poultry and lamb by the burgeoning Muslim population of this country and now about 40% of poultry and 25 to 30% of lamb is now slaughtered to meet the Halal criteria. In 2012 the FSA revealed that 80% of animals slaughtered as Halal are in fact stunned. However all animals slaughtered as Kosher for food for the Jewish population are not stunned.

European Council Regulation 1099/2009, which was published on the 18 November 2009, requires stunning before slaughter, but allows member states to apply an exemption for religious slaughter. Each member state can refuse to exempt religious slaughter from stunning regulations. Would you believe that our pathetic politicians have once again rolled over and refused the option of humanitarian common sense just in case they will offend some barbaric religious practice? What has happened to this once lovely country of ours which used to stand up for decent values, amongst which was the welfare as far as possible of dumb animals, which give up their lives for us to eat? They should be treated with humanity and respect right up to their deaths for our benefit, not tortured in their dying moments.

Apparently in 2012 a Right Honourable MP and Minister of State for Agriculture and Food was known to have said that the killing of food animals without stunning is not acceptable in the Western World. So far so good, but he then went on to say that we, I suppose that must mean the original British, need to be tolerant and understanding of religious communities who for religious reasons want their meat produced in this way. No such understanding seems to be apparent for Westerners and their traditional pursuits in particularly the Muslim countries.

At that point he ceased to be "so-called" honourable and became a lap-dog to political expediency, on the one hand saying that it is totally unacceptable but is prepared to allow the barbaric torture to continue; even his boss that two faced Cameron had indicated in 2011 that there was no question of a ban. He went on to say that the Jewish community has been a good example in integrating into British life but that does not mean that you have to give up things that you hold very dear in your religion.

I would say to Cameron, if integration involves the breaking of the law of the country into which you are integrating in the name of religion or anything else for that matter then the law (made by politicians) has been made to look an ass, together with the politicians who say one thing and then do another. They deserve nothing less than total contempt by those citizens whom they purport to represent.

In 2012 Professor Bill Reilly a previous president of the British Veterinary Association, while writing in the Veterinary Record said that in his view the current situation is not acceptable and, if we cannot eliminate non-stunning, we need to keep it to an absolute minimum. Surely however commendable his observations, it rather misses the point. Our European masters have said that any member state has the right to ban non-stunning for whatever reason "full stop". One animal tortured in this despicable way is one too many and decent thinking countries in Europe have already banned it, it begs the question why, if people like a Minister of State and Professor Bill Reilly agree that it is totally wrong why do we not just ban it? It does not seem to have brought the decent countries that have banned it crashing down. I for one would have a tiny bit more respect for our political class it they had the courage to uphold our laws. Personally I have tremendous respect for the Jewish people who have suffered terrible persecution for countless years particularly during the last war. They have risen out of all that tragedy and today in their homeland of Israel they stand for no slight from other countries and if threatened or attacked will retaliate with great courage. Why I would ask, if the other faith which carries out ritual slaughter of food animals can accommodate, at least in part, the stunning prior to cutting of our food animals, the Jewish faith which has known so much by way of suffering cannot accept a humanitarian point of view when it comes to their food animals. The Kosher was a system put in place back in the mists of history for the very good reason of forbidding the consumption of carrion, but surely we have moved on from there.

The other factor of course is the lack of recognition or identification of the flesh of animals that have been treated in this barbaric manner. To save expensive differentiation between Kosher, Halal, and traditionally slaughtered meat, big business chooses not to indentify the different categories, which when you consider how stringent labelling is in other directions it seems something of an anomaly. As with the labelling of horse flesh, which is and always has been a legal requirement, I believe all ritually slaughtered meat should be adequately labelled so that the non-religious consumer can boycott or not what they may find offensive on cruelty grounds.

The clear labelling of meat obtained from animals and poultry which have suffered unnecessary agony during their dying moments is the obvious answer to all this. Various groups have a vested interest in not labelling it including the religious groups for religious reasons and the abattoirs that accommodate these practices for financial reasons. The big wholesale meat suppliers to hospitals, schools etc. for financial reasons but also to avoid the complication of trying to supply both unstunned ritually slaughtered meats together with that which has been humanely slaughtered according to the law of the land.

It is pointless to hope that our political class will do the humane and decent thing, even though hopefully the vast majority of our citizens who put them in a position of power, something I think they very often forget, would wish they would do so. They have a vested interest in not appearing to offend any religious groups and if history is anything to go by they probably have a financial interest as well.

Going back forty odd years all this was a problem then. As I said earlier in this chapter, Kosher hind quarters of beef were commonly on sale in the meat depots situated in Old Market Street in Bristol and people would unsuspectingly buy this unidentified meat from their butcher's shops. The killing of young goats and poultry was commonly practiced in the back gardens around the St Paul's area of the city.

Les Mawditt and I did our best to control this illegal practice, helped by a certain amount of local knowledge. Today, with the huge increase in the ethnic populations, I am sure this practice continues, but without the will of officialdom who are also frightened to death of offending some religious group or other nothing will be done about it. Unfortunately, some people think they only have to play the race or religious card, and then they can do as they like, regardless of the law of the land. Sadly they are probably right.

CHAPTER 25

CASUALTIES

OVER THE YEARS I have been involved with many, many food animals which have been considered to be casualties. These animals fall into two broad categories: first those that have become injured in some way which necessitates emergency slaughter; usually this category involves broken bones, mostly legs. The flesh from these animals can often be salvaged for human consumption other than those areas immediately adjacent to the injury which are often damaged or bruised. Secondly, those animals which are suffering from some disease or condition which precludes them from being sold in the normal way, either through the market process or else directly to the abattoir. The more devious farmers, of which there are always some, will try to sneak these sickly animals into batches of otherwise healthy companions. The more responsible and feeling farmers, who in my experience form the vast majority, will bring the individual animal in on its own as a casualty.

The rules have changed several times over the years as to what you have to do when bringing such an animal into the abattoir. At one time you could just bring it in. Then came self-declaration certificates, which the owner sent with the animal declaring his diagnosis of the condition concerned, and also what if any drugs had been administered to the animal, and when. This was all to do with the withdrawal period prior to slaughter. The owner was put on his or her honour to sign and declare when various drugs had been administered in an effort to cure the animal's condition. We then moved into the realms of veterinary certificates, not only declaring as to what drugs had been administered, but also the vet's opinion as to whether the animal was potentially fit for human consumption. Now of course, since we have been in Europe, if a farmer has what he deems to be a casualty animal he must call his vet to assess the animal concerned, not only for drug administration but also for his or her opinion of the condition and as to whether the animal is fit to be travelled to the abattoir. If not, a licensed slaughterman has to be called to stun and bleed the animal, and once dead it can only be transported over a limited distance to an abattoir willing and able to accept casualties, which is not always the case. Arrangement also has to be made for the resident OV to be present at the abattoir to assess the animal again prior to acceptance

When all these latter arrangements are put in place the cost to the owner can be so prohibitive that the returns for any salvageable meat (which may or may not be the case when presented to the meat inspector) make it in many instances a non-starter, and farmers tend to just ring up the knacker man or hunt kennels to kill the animal and take it away. This is fine in many cases, but in other cases potential food is yet again is being wasted.

Casualty animals have now not only to be inspected by the OV at ante-mortem but also post-mortem, and not by the meat inspector. It is of course alright for the meat inspector to examine the general run of animals coming through the plant and detect ones that have a problem of some sort, but when it comes to an animal which is already known to have a problem, and what that problem may be, the meat inspector is no longer good enough to make a decision. In practice most young OVs leave the inspection of casualties to the meat inspector, but legally it still remains the prerogative of the OV.

All this was put in place by people behind desks who do not have an inkling of what they are presiding over. Yet again, you couldn't make it up.

Out of the hundreds, if not thousands, of casualties I have dealt with over the years I would like to pick out just a few, to illustrate what a huge variety of interesting cases we as meat inspectors see on a fairly regular basis. I will start with something which used to be a very common occurrence, and which is a condition of intensively produced pigs. Pigs are probably the most intelligent of our food animals. When kept intensively they have nothing in their miserable lives to keep them occupied. Bear in mind that naturally they would be living primarily in woodland, as do the wild boar roaming various parts of our country today, and they would, when not sleeping or mud bathing, be rooting around their environment looking for food items such as worms, beetles etc, as well as natural fruit such as blackberries and acorns. When confined in square bland pens with nothing to keep them occupied these inquisitive animals are continually looking for something to do between feeds. The curly tails of their fellow inmates can attract their attention and they can sometimes start to chew at them. This condition is known to us in the abattoir as "tail bite". Once the tail skin has been perforated the way is open for the entry of bacteria. The condition would manifest itself by a red swollen infected tail through to, in a bad case, a large black suppurating cavity in the pig's back where its tail used to be. Sometimes the pig would become paralysed when the infection had spread down the spine in the visible form of abscesses, occasionally and intermittently right down to the neck area. At any stage in this process the animal could become a casualty and was rarely salvageable, suffering from a condition known as "tail bite pyaemia".

The law requires that the pig is split in half down the spine for the inspector to examine the spine on both sides for evidence of abscesses. When teaching students I tell them that the first thing to do before having the pig split is to examine the lungs and occasionally the kidneys for evidence of abscesses. If abscesses are found, even one, I will always reject the pig. The only way that the infection can have spread from the infected tail is via the bloodstream, and therefore the condition is what we know as systemic. By not splitting the pig contamination of the equipment will be kept to a minimum.

Sadly many inspectors today have not the time on the fast-moving lines, doing thousands of pigs every day, to closely monitor a tiny abscess in the lung and connect it with what appears to be a not very badly infected tail, and all that is done is maybe remove the tail or at best have the pig split, which may not reveal any abscesses in the spine. Sometimes it is difficult to explain to an abattoir owner or a pig farmer the intricacies of bacterial spread when all you are showing them is a slightly swollen tail and a less than pea-sized abscess in the lung.

Going right back to my days at Gordon Road, occasionally a knacker man working in Bristol would come in with his cattle lorry having picked up an animal which he thought might be good enough for human consumption. He would ask one or other of the inspectors to have a look at it to see if it was worth unloading for slaughter. On one particular day something occurred which hopefully would never happen today, although I have heard rumours within the last 10 years, but fortunately nothing that concerned me personally. The knacker man brought in a young Aberdeen Angus steer. It was my turn to go and have a look at it and give my opinion as to whether it might be passed as fit for human food or was not worth unloading and should continue its journey to the knacker yard and be used for dog food. Much was hanging on my judgement, because if I said give it a chance and then it was subsequently condemned and thrown in a bin the knacker man would lose whatever it might have been worth for dog food, but conversely, if it was passed for human consumption, considerable profit could be made. The animal concerned on this day was lying down in the front of the knacker lorry. Between me and the steer were two cows which were already dead, had been collected, and would obviously only been used for dog meat. I climbed up into the lorry and made my way over the dead cows to assess the steer concerned. It had been lying down quietly up until this point in the proceedings. Obviously alarmed at a white-clad figure approaching it, it quickly rose to its feet and charged. I had nowhere to go and, fortunately for me, but unfortunately for the steer (which it turned out had a broken back leg much like the one pictured), being badly incapacitated as it was, it fell over the two dead cows. It immediately tried with difficulty to regain its three legs which allowed me enough time to escape the lorry, but not enough time to close the gates on the back. The steer half rolled and half walked down the gangplank and then, having regained its three legs, proceeded to try and kill anybody stupid enough to stand in its way. When you see an animal which is absolutely terrified, and in this case obviously in extreme pain, you cannot help but admire its courage against what it perceives as its tormentors. The only kind thing to do is to shoot it as quickly as possible to relieve both pain and stress, which is of course what we did. As far as I can remember the vast majority of it was passed as fit to eat, but any animal which is stressed at the point of slaughter will often cut dark in the muscles and is probably best used for manufacturing purposes.

During the last couple of years I did hear of an OV being killed by a bovine animal while doing ante-mortem inspection of some cattle being unloaded from a lorry at an abattoir. These large beasts can at any time be very dangerous. Our so-called European partners in Spain, love to torment black bulls by running them through the local streets on the way to their final torture and death in the bullring. I find it interesting how many mindless yobs full of testosterone get run down and gored by the terrified bulls. Incidentally, I did hear that the flesh from these fighting bulls is traditionally given to the poor, that was probably before any European legislation came to bear on the tradition. It is probably now is sold through a devious route to supermarkets to put in their processed meat products, together with the horsemeat, but I digress.

It is easy to see then how powerful and potentially dangerous these large bovine animals are, when every year people are killed or injured, usually while walking their dogs on farmland where cows and calves are grazing. Instinctively, cows will protect their calves from what they perceive as predators.

During the lambing season at Gordon Road, where we specialised in the slaughter of ewes and rams, we would inevitably get many casualties which either were unable to lamb or else had lambed and been damaged and/or infected as a result. There is a condition known as "twin-lamb disease". This would affect ewes which would be carrying two or more foetuses. This was fine until maybe there had been a reduction of nutrition for the ewe, or maybe the multiple pregnancy was more than her body could stand. What apparently was happening was that the foetuses were draining their mother's body of nutrition faster than she could replace it. The ante-mortem symptom was usually a recumbent animal which was completely out of it, just lying there with a glazed expression on her face. She was usually very close to lambing when this condition occurred and the lambs inside her were often very much alive and well. The farmers concerned often had other very recently lambed ewes at home and were very often able to foster another lamb onto a ewe which had only had a single lamb. To this end I became very proficient at performing caesarean sections on the casualty ewes so that their entire pregnancy would not have been for nothing. Often with the assistance of one or more of the slaughtermen we would gently lift the unfortunate ewe onto a legging cradle. When in position she would be shot and bled in the normal way, and with the boys holding her back legs apart I would carefully open first her belly and then her uterus, taking great care not to cut the lambs inside. One of the boys would tie a piece of string tightly onto each umbilical cord to minimise haemorrhage when they were cut, then I would put the lambs on a sheep skin and rub them down with some tissue and make sure that their airways were clear, and that they were breathing normally. This was best achieved by holding them by the back legs and swinging them gently around upside down to clear the amniotic fluid from their lungs. Making sure they were well into this wicked world I would return them to the farmer together with as much colostrum (first milk) as possible in a beaker for the farmer to take with him to give to them in a bottle when he returned home. This contains all the mother's antibodies to give them the best start in life against all the bacteria in their particular environment. Hopefully in due course they would be fostered onto a surrogate mother.

I remember some years ago I was working my cocker spaniels on a lovely friendly farm shoot down in Somerset. The owner farmed very high pedigree Suffolk ewes for the production of Suffolk ram lambs for future breeding stock. These pedigree sheep were of considerable value to their owner and he was very proud of them. After one particular day's shooting, when everyone had retired to his shooting lodge for tea and cakes and a small quantity of alcohol, I noticed that the farmer kept leaving the room and going out into his lambing shed which lay just behind the lodge. When he returned after several occasions he appeared to be increasingly worried. I asked him what seemed to be his problem. He said that one of his Suffolk ewes had been trying to lamb, but he was unable to turn the first lamb which was apparently twisted up in the uterus. He was a very experienced farmer and I knew that if he could not lamb the ewe then the only hope was for a caesarean section, which would have to be carried out by a vet. It was by now late at night and the farmer could not raise his vet for whatever reason. I told him that, to salvage something from the sad situation, because he did not think that the ewe would live much longer, I could perform a caesarean section to try and save the lambs, but that it would be terminal for the ewe. He agreed that if I could, that would be the way he would like to go.

Using the farmer's .22 rifle and a folding knife which I have always kept in my pocket issued to me many years ago when working for Bristol, to use when inspecting meat away from the abattoir around the city, I enlisted the assistance of several others at the shoot with hopefully fairly strong stomachs. I requested a nice clean bed of straw, some hot water and some fairly fine string. I went through the procedure well rehearsed at Gordon Road and produced two huge Suffolk ram lambs. As soon as they were able I let them suck on their dead mother's udder to obtain as much colostrum as possible before being fostered onto other ewes. Returning to the carcass I dressed it and, with the aid of others, hung it with a piece of rope threaded through its back hocks and thence up onto a roof beam. I said that I would return in a couple of days and butcher the carcass, which with long and careful cooking as mutton would be very nice. The two lambs survived and grew up. The mother's carcass made many meals so we had turned what could have been a complete disaster into something not quite so bad. I think that I even sold the skin to the hide merchant. My friend the farmer was very grateful for my efforts. It is a pity that there are not more people around today who are able to do similar things to save life, and also save the waste of potential food.

Ewes were one thing, but one day a cow came in which could not calve and the farmer (probably one for whom I had saved some lambs on another occasion) asked me if I could save the calf out of this cow. Having never before done a caesarean on a cow I could see all the possible implications, like getting kicked to death by the reflex actions of a dead cow. However, I said I would give it a go. The cow was shot in the normal way and rolled out of the stunning trap onto the floor and bled, whereupon her top back leg was shackled and the strain of the body's weight taken on it. This put one back leg out of action and the other back leg was tightly held by two slaughtermen. I dived into the middle of all this and dropped the udder out of the way so that I could incise the belly and subsequently the uterine wall. All this had to be done very quickly. The calf was pulled out and, having tied off the umbilical cord, deposited on a pile of waiting sheep skins. I cleared its airways and rubbed it down as best I could, something the mother would have done with her tongue. I then removed the udder and told the farmer to let the calf suck as much of the first milk as possible from it. I heard afterwards that the calf lived and subsequently thrived.

One day at Nailsea a local farmer, who keeps quite a lot of horses for breeding purposes, came in and asked if I would try and save a foal, because the mother was dying and the vet could not get there for some time. Grabbing all that I thought I would need for the job I returned with the farmer to his farm. There were several strong helping hands to hold the back legs apart and I shot the mare. As I began operations I had completely forgotten to cut the mare's throat (which reduces blood pressure), knowing that she was only going to the hunt kennels. As soon as I started I realised my mistake, and as time was of the essence to withdraw the foal and there was nobody there capable of cutting the throat, I had to go ahead regardless. Fortunately the guys holding the legs were very strong and as this was a very large mare were saving me from being kicked to death. I managed to cut my way into the foal but in doing so had become covered in blood from head to foot by the spurting blood vessels full of pressure which I had severed on the way in. However, despite my personal discomfort and that of those who had helped me we birthed a healthy foal.

Fortunately blood and sometimes stomach contents and faeces are easily removed by soap and water. I have often thought during my working life that I would rather mess around with the insides of cows rather than mess around with insides of cars as some have to – oil and grease are so difficult to remove.

When we arrived at Gordon Road on Monday mornings the first job of the day was to inspect any casualty animals which had been killed during Saturday afternoon (we used to work on Saturday mornings) and Sunday. These animals would have been killed by one, sometimes two slaughtermen, who would be on call during this period of closure, as nobody can predict when a casualty may occur. The resultant carcasses would mostly be of cattle, but there would be some sheep and very rarely pigs, which would have had to be skinned because there would be no scalding facilities available over the weekend. Each would be hung up with its red offal by its side and the green offal on the floor below, for identification and correlation. In those days there was no requirement for the animals to be ante-mortem inspected as there is today. This was when your knowledge and experience was called into play, not unlike a human pathologist looking at the cause of human death from disease etc. We are food animal pathologists with the added responsibility of deciding whether or not the animal is fit for human consumption. There was sometimes a bit of history involved, and sometimes the reason for emergency slaughter was self-evident in the form of broken limbs etc. But sometimes a bit of head scratching was necessary before an important decision could be made.

Today the huge conveyor belt meat factories have to operate at terrific speed where all the inspectors can do is put a colour-coded ticket upon a carcass they do not like the look of, and the carcasses concerned are withdrawn at a further point down the line and placed in a detained room. A different inspector has to assess what needs to be done to the carcass, having never had the entire carcass and offals to look at together. All he has is a colour-coded ticket to guide him with regard to the disease or condition concerned. Personally, I like to have sufficient time to assess both offal and carcass together before making my decision.

I once worked for two days at a large pig abattoir. I can with my experience inspect as fast as any, but I felt on the first day under pressure from the sheer numbers of animals coming past me on a moving line; I may add that there were several inspectors working on each part of the line. On the second day the senior at the time asked me to work in the detained room to experience that part of the inspection procedure at this factory. This happened only about 10 or 12 years ago, after I had ostensibly retired and was only a casual inspector for the Meat Hygiene Service. I told the senior inspector that I was not prepared to try and do my job under the circumstances at that factory, one among many such factories out there today. I felt that the working practices did not allow me to do my job as I would have wished. Happily, I was in the position of being able to refuse work if I so wished, and I was not prepared to be rushed into doing what I had at the time been doing under happier circumstances for 40-odd years. I fully expected not to be asked to work again, but very soon after this event I was asked to work at another, again big plant, on cattle this time, where sufficient time was given to do the job properly and I was again happy in my work.

Other full-time inspectors working in these factories today, many of them European vets with little experience, had no choice in the matter if they wished to continue their employment with one of the agencies contracted to the Food Standards Agency. I feel very sorry for them but maybe due to their lack of experience they cannot see a problem.

Pointing out the money-orientated service in place today will not endear me to the powers-that-be. Fortunately, I am in a position to voice my reservations, which so many other good experienced inspectors with whom I have spoken agree with, but who cannot for employment reasons say so in public. I speak on behalf of many disillusioned inspectors out there. Like so many things in life today, when money and big business are involved, honesty and trustworthy behaviour somehow become compromised. But of course those who should be of the highest possible moral fibre call themselves honourable gentlemen.

Returning to the casualty animal, sometimes animals which do not really come into the casualty category are presented. I remember one day a four or five-month-old shorthorn calf came in. It had five legs:

the additional limb hung down uselessly from one of its shoulders but in no way did it impede or debilitate the young animal and therefore it could easily have been left alive, but as it had been accepted as a casualty the farmer didn't want it and it had to be despatched. I have many birth abnormalities in my collection of morbid specimens. Some are pickled in formaldehyde and some are dry-bone specimens. One in particular is an almost complete double calf: two heads, four front legs, two umbilical cords, two tails, two fundamental orifices, but only two back legs. This very unusual deformity was far too big for a glass tank, and I paid to have it embalmed by a very interesting and capable young man who hailed from Newport in South Wales. In his everyday life he embalmed human cadavers, and he told me that he had on occasion worked on young soldiers who had been blown to pieces in Iraq and Afghanistan, trying to put their mangled bodies back together as far as possible so that their relatives could view them in death. They were sent to war by Prime Minister Blair and his cronies. He is now, grinning like a Cheshire cat, going around the world under the misnomer of Peace Envoy. War criminal would surely be a better description. This young embalmer sewed the back legs back onto my calf, as they had been removed by the veterinary team who had come from Langford initially to try to calve the cow concerned. Finding that they could not calve the cow in the normal way they decided to cut up the calf via the vagina, but having removed the back legs in this way, they then changed their minds and decided to do a caesarean section to remove the rest.

These deformities cause much interest amongst my students and the various groups of lay people to whom I give talks, from Rotary members through amateur historians to Young Farmers and Masonic Lodges. My subject, which is normally kept behind closed doors, is sometimes an eye-opener for normal people.

Another probably unique specimen is a sheep's back leg which presented with a large swelling in the back of the thigh. An incision was made by the inspectors in a Wolverhampton abattoir and a complete foetus was found. A friend who presently works at Langford and was one of the meat inspector students with me at Langford about 15 years ago brought this incredible specimen of a parasitic twin down to me as a present. He worked for a while at abattoirs in the Midlands but for some reason he never forgot his training period at Langford. He was not alone in his love of the Langford campus. Another, who has also become a good friend, worked for a while as an inspector in the Wiltshire area but eventually came back to Langford and now works and lectures there himself. He has also written several books on the subject. His wife, whom he met on the same course, went on to work as an inspector around the Gloucestershire area, but ultimately she felt the need to return to Langford and she also now works there herself. There must be something about the place!

A further, probably unique, specimen was found for me at a huge pig abattoir near Bristol. This particular animal was in possession of two hearts. It was found for me by yet another past student, a young woman who was with me some years ago and who went on to become a very proficient inspector in her own right. As far as I know she has not yet applied for work at Langford. These people were not alone in remembering my passion for collecting meat inspection specimens, and over the years erstwhile students have sent me interesting items from all around the country.

The man who now lectures himself at Langford has started a meat inspection museum with many of his own specimens, which he hopes one day to embellish with my collection when I die. He also suggested that perhaps when I die, and as I have been going there for so many years, I could be pickled in a tank of formaldehyde and put on display for the next generations of students to see! I must admit that the thought appeals to me, as I have a morbid sense of humour. Taking the idea a step further, even though I was fully aware that he was joking, I knew that another friend of mine at Taunton, whom I see during the shooting season, worked for a company which made reinforced glass used today as flooring and for putting over the

American Grand Canyon and other places for tourists to walk on and look down into the abyss. I asked him if a coffin to fit me could be made out of his glass. He said that of course it could, but perhaps a stainless steel coffin with a glass lid might be better, with a cost of around £600. The whole could then be inserted under the floor with just the glass lid to walk over. Armed with this information I returned to my friend at Langford to give him the good news. A look of horror came over his face, and he told me he had only been joking. I told him that it was too late and that I had ordered the coffin, but he said that the university would never allow it. Sadly something which appealed to me and I thought quite appropriate was never to be and I told my friend at Taunton that it had all been a fairly elaborate joke, and that I would no longer be needing the coffin.

Two years ago I had to have a hip replacement operation which was done at Weston-super-Mare Hospital. When on a pre-operational visit to the surgeon I asked whether I could have my old hip joint back for training purposes with my students. I have in my collection many examples of arthritic hips from cattle, sheep and pigs, but a human one would have been very good from a comparative anatomy point of view. The surgeon was very reticent and observed that bureaucracy being what it is it would be very difficult with so many eyes watching in the operating theatre, but he would see what he could do.

On the day of the operation I elected to have an epidural anaesthetic in preference to a general. Incidentally, for those who have not experienced it, the feeling in your lower body is very peculiar. You have no mobility and no pain and your legs and feet do not respond to brain signals at all. Sadly probably the lack of feeling experienced with an epidural is the same as that experienced by those who for whatever reason have tragically become paralysed in their lower body.

I was wheeled into the operating theatre on a trolley and was totally aware of what was going on around me. I was hoping to watch the procedure as far as possible, but a sheet was put over me on the side that was going to be operated on. The best I could do was ask the young anaesthetist sitting by my head for a running commentary of my operation, which he did very well. At the end of the operation I asked, while still in the theatre, what had become of my old hip. I was told that it had been thrown in the clinical waste bin. I asked for it to be retrieved, but apparently this is not allowed. I was very disappointed because I would only have one, possibly two, chances to obtain a human arthritic hip.

The next morning in the ward a young intern came to apologise to me, but he said that with so many eyes watching they dare not comply with my request. I pointed out that at the end of the day it was in fact my property and had been taken from me against my wishes. Later, when on a follow-up visit, I took in my pocket some good pictures of arthritis in cattle, sheep and pigs together with a clean bone specimen of a sheep's pelvis and femur with extreme chronic arthritis in my pocket. I showed my surgeon the pictures and then produced the femur and pelvis from my pocket and asked him to rate how bad was mine. He laughed and said "Nothing like as bad as that." I gave him the pictures as a present and said "You can see now how much I wanted my hip back." He apologised again for the madness of the bureaucrats but said the ruling was out of his jurisdiction and that he would have liked to have complied with my request. He could see now how complementary my hip would have been amongst my collection of very similar food animal problems.

Fortunately for me there was a very successful treatment available for my hip; I have also seen this procedure practised on pet animals, but for food animals the terminal solution is by far the kindest. If you allow your pet animal to suffer from some incurable condition you lay yourself open to prosecution for cruelty. The same cannot be said for those humans who are allowed to suffer miserably for sometimes very long periods when the final end is inevitable. Surely in the name of humanity these people who have just had enough can be allowed to die in a humane manner when it is their wish rather than to have to suffer for months, sometimes years, before the inevitable occurs.

It is interesting that those who oppose a dignified end are not the ones who are suffering, or who for some religious reason feel the suffering has to continue. Inevitably protection must be considered for those individuals concerned, but a panel of doctors could be involved to consider each potential case on its merits. By reducing the population and freeing up many, many hospital beds the cash-strapped National Health Service would not be wasting resources on hopeless cases who in many instances do not want to go on in pain and suffering anyway. The so-called Liverpool pathway from what I understand is cruel in the extreme. Denying anything, be it human or animal, food and water is tantamount to torture, and should be condemned at every level, but a tablet to let those poor individuals who have done their allotted time and are now in extreme distress slip painlessly and peacefully into death can only be good news for all concerned. This would of course relieve the pain of the relatives who in many cases do not want to stand helplessly by and watch their loved ones sink slowly and often painfully into death.

When working at the old Nailsea in 1981, even though the abattoir was comparatively very small, we still had our share of interesting casualties. Casualty pigs from a large intensive unit were brought to Nailsea on a weekly basis in a horsebox. I would be asked to ante-mortem these poor animals by the farm manager, and would then decide which individuals were worth slaughtering in the abattoir and which stood no chance of being passed for human consumption. I would shoot all those which came into the latter category and they would be put in a waste bin, without potentially contaminating the abattoir itself. All these animals came under the general heading of percentage loss to the business. They came in all sizes, from little weaners right through to pigs almost ready for bacon.

One day an animal which could apparently not eat or regurgitate its cud was found to have a potato wedged halfway down its oesophagus. It had obviously been fed on stockfeed potatoes; unfortunately they had not been properly crushed or shredded and this potato was just the right size to cause this blockage. This unfortunate condition is not unknown to the veterinary profession and I understand there is a piece of veterinary equipment available, shaped like a long corkscrew, which can be introduced down the oesophagus and screwed into the potato which can then be withdrawn.

On another occasion a bovine casualty brought in by its owner was shown to me in the lairage. I was told that the animal was becoming listless and losing condition. I could see as the animal turned its head from one side to the other a slight swelling on each side of its neck, just below the back of the jaw. I said to the farmer that in my opinion the animal had TB and that the swellings that could be seen were the grossly enlarged sub-maxillary lymph glands lying just under and inside the back of the jaw. These are the glands that your GP feels for when you are suffering from a bad sore throat and other more generalised problems. The farmer said that he did not have TB on the farm at the time and he was very sceptical about my diagnosis. With the benefit of hindsight on post-mortem examination TB was indeed found in both sub-maxillary lymph glands and also elsewhere around the body, which necessitated the total rejection of the carcass and offals. When talking to the farmer later he was magnanimous enough to say that I must be gifted with second sight.

One day just before Christmas, when working at the old Nailsea abattoir, a car with a tiny trailer behind came in. I was asked to go and have a look at the animal in the trailer. The owner was standing by the side of the trailer, and as I approached he began chucking the ewe (which is what the animal turned out to be) under the chin. She was sitting upright in the tiny trailer with her head bolt upright pointing forward just like a figurehead on an old sailing ship. The owner's hand movement under her chin was making her head bob up and down. I could not believe what I was seeing and stopped to watch for a second or two. It was blatantly obvious that the ewe was stone dead and had been that way for quite a while because it was stiff as a board sitting in the upright position. I asked the owner what he thought he was doing; he replied that he was comforting her prior to her death. But I said "She is already dead, and has evidently been that way for some time." "Is she?" he asked. "She was all right just now!" To this day I really do not know whether the owner thought that I had come up with the grass, or whether he really thought the ewe was still alive. I said "Come on, take her away to the hunt kennel; we *never* accept an animal in a moribund condition at the abattoir."

Another animal which came into Nailsea had an intermittent walking problem: it would take a few steps and then sway from side to side almost collapsing, but just able to regain its balance to proceed further. Again with the benefit of hindsight after slaughter I found that the animal was suffering from a curvature of the spine and had grown a spur of bone inside its spine which would intermittently pinch its spinal cord, causing intermittent paralysis.

Later at Nailsea I developed a system of my own which involved the farmer with a problem animal needing slaughter ringing me and describing over the phone the problem that the animal presented before bringing it to the abattoir. From the farmer's description I could generally diagnose the disease or condition from which the unfortunate animal was suffering. Some of these were definitely non-starters from an eating point of view, and probably in any case would potentially contaminate the abattoir if slaughtered within the premises. Conditions such as that of a calf with swollen leg joints which probably could not even walk; this is a condition known colloquially as "joint ill", properly known as umbilical pyaemia. This is an infection which gains access to the body via the not-yet-healed umbilical cord, which becomes swollen, and the condition at that stage becomes known as "navel ill". The organism travels quickly up to the liver and often will kill the animal at this stage of the infection. However, if the animal survives this initial onslaught, the organism gets into the bloodstream and sets up home in one or more of the joints, causing extreme pain. Because the organism is now a systemic infection there is no way the animal can ever be passed for human food. I would tell the farmer over the phone that we would not accept it at the abattoir and his best option would be the knacker man or possibly the local hunt kennels.

Another possibility would be an animal which had slipped and fallen, maybe on a slippery floor in the milking parlour. This could result in another colloquially termed condition where the animal concerned would be deemed by the farmer to have "spread" itself. The animal would be recumbent or possibly trying to drag itself about with its front legs, with its back legs spread-eagled behind. This "spread" when examined at post-mortem can be one of three possibilities: the first is the dislocation of the hip, the second a splitting of the pelvis, and the third a severe tearing of muscle away from the pelvis. All three can cause major distress, and the sooner the animal is dead the better. To save putting the animal through the extra distress of loading it in a trailer, I would tell the farmer to leave it exactly where it had fallen and I would be with him as soon as possible to shoot and bleed it and thereafter he could load it upon a flat-bed trailer, cover it with a tarpaulin and be with us at the abattoir as soon as possible. Much of the resultant flesh would be useable and the animal would be saved from the additional pain and stress of being loaded and brought in alive.

This happy relationship with the local farmers could only be enacted in small abattoirs with the facility to accept and hoist recumbent large animals, and only because I have always maintained my slaughtering licence, putting me in a much better position to criticise on occasions and guide on other occasions slaughtermen with whom I have worked over many years. This situation could not be replicated today, bearing in mind that very few if any meat inspectors hold a slaughtering licence, and even if they did they may not now assist the "FBO" (Food Business Operator).

Have you noticed how everything today is reduced to abbreviations? I remember one senior officer, when I was working in an abattoir on the Bristol fringe, addressing me with a sentence completely composed of abbreviations and acronyms. I had no idea what he was talking about and told him that should he wish to communicate with me in the future then perhaps he could use the English language.

To return to not assisting in any way, what the bureaucrats fail to understand is that the FBO should not be regarded as the enemy, as some of them seem to think, and that working as we do for the Food Standards Agency amongst employees of a private company it is essential that a working environment of mutual understanding and camaraderie should prevail, with a combined effort to put a very high standard product out of the door. This horrible feeling of mistrust, encouraged by the people in power sitting behind desks at the top of these agencies, who often have little or no knowledge of the practicalities of the actual job which they oversee, promotes a "them and us" environment with potential prosecution of the FBOs as a first resort rather than, as it should be, a last resort.

During my career I have seen a number of sheep brought in as casualties which have been "worried" by domestic dogs. Sadly this situation is largely a result of the lack of knowledge and understanding of the natural world by the ever-increasing urban population. People go into the countryside for sheer pleasure and recreation, and rightly so. Many take their domestic household pet with them for a long walk. So far so good. What they do not understand is that they are taking a genetically programmed predator into a field of food. The domestic dog, an animal which I have loved over the years, is in fact a tame wolf with all the instincts that have evolved over millennia. Unless the owner has taken the time and trouble to curb those natural instincts, accidents will happen. I and many others have channelled those instincts into something useful, as with working gundogs which have been well trained, and also collie dogs, where their killing instincts have been harnessed into helping man manage his domesticated farm animals. To slightly digress, people say that foxes are cruel because they will kill all the poultry in a chicken house. There is nothing cruel in nature: the fox is only responding to instinct, which is that when food is available they should grab it as it may not be there tomorrow. Man is the only species to my knowledge who will inflict pain or death for the sake of it, often in the name of religion.

To return to the casualty sheep being brought to the abattoir, they will often show signs of debility, and sometimes some blood will be visible around the head and neck area. I did see one where the dog had skinned and probably eaten flesh and skin from the entire head and the sheep had come in still very much alive.

In my experience, what is visible at ante-mortem on the outside dramatically hides what is in fact horrendous damage under the skin. The way that the dog (wolf) works when killing, often an animal larger than itself, is to try and debilitate it as much as possible with multiple bites which will weaken it and eventually allow the smaller dog to pull it down and kill it, often by suffocation on the throat area. Dogs (wolves) are equipped by their cleverly designed dentition to inflict terrible damage to their potential prey: this can be seen by the damage caused by a comparatively small dog to us as human beings. A sheep which was been "worried" very often has to be rejected anyway for multiple bites and general bruising. There is also a school of thought which says that the intrusion of a dog's teeth into the bloodstream of the sheep automatically necessities total rejection because of the risk of bacterial contamination from the dog's saliva.

Sheep are not the only casualties of the lack of knowledge and often arrogance of people from an urban background when walking in the countryside. Footpaths are historically only wide enough for people to walk in comfort. Technically people and their dogs should not stray past these limitations; and unless their dogs are very, very obedient they should be kept on a lead at all times.

In the immediate vicinity of my home there is a long and fairly wide area of old woodland. Many people, who have moved to the area probably from more urban situations, walk this woodland with their dogs. Frequently barking dogs can be heard travelling through these woods completely away from their owners, coursing roe deer through the trees. Usually the adult deer can outpace these dogs, but the little fawns in the springtime fall easy prey. This year one half-grown roe fawn was driven by dogs over a quarry face nearby; it scrabbled to hang on to the undergrowth at the top of the cliff-face screaming in fear, but eventually fell to its death below. Another adult, this time about half a mile from me, was cornered in an orchard. It was still alive when shot, but the resultant carcass was deeply bitten all over. Yet another was thought to have been a road traffic accident, but was in fact debilitated by TB and was pulled down by dogs. The problem is continually increasing, with suburbia spreading further and further into the countryside. The number of people with lack of natural knowledge is also increasing year by year. It has to be said that there are those who go about their country visits with a high degree of responsibility and are fully aware that their loveable pooches can turn in an instant into lethal killers.

One day a man was walking his two fairly young Alsatian-type dogs through these woods, and they were totally out of control, crossing an intervening field between our property and the woods. They saw my Soay sheep in our field. Being young dogs they started to chase them as a game. Fortunately I just happened to see them before any damage was done. I managed to catch one of them and put it on a slip-lead, possibly in retrospect a silly thing to do not knowing whether the dog would attack me. I shouted to the man, who was still in the wood, that I had his dog, he came across with the other dog which had returned to him by then. He was very abusive and kept saying "Let my f...ing dog go!" I told him that he could recover it from the police later having already told my wife to ring them. I dragged his dog half walking and half on its back down the field and locked it in one of our barns. The man had illegally entered our property over the fence and had followed me down. I stood outside the barn with my arms folded for two hours, waiting for the police to arrive. The man was intermittently abusive and kept wanting to know the time, because he said he needed to take his tablets. My rejoinder was that I couldn't care less about him or his tablets, and hopefully he might die for a lack of taking them in time. When eventually the police arrived I let him go first with his arrogant explanation. The police officer listened carefully; I then told the man that I would have been perfectly within my rights to have shot his dogs, with which the police officer agreed. I would never shoot a dog under these circumstances, it is not the dog's fault, and it is always the owner's fault.

On another day I was lucky to see a little slightly overweight spaniel chasing my sheep, this dog had no chance of catching one, but was trying hard. I called the dog to me and locked it in one of my kennels. The owner's telephone number was on its collar. By now I could hear a voice in the woods frantically calling for a dog. I let this go on for about half an hour, and then rang the number. "Have you lost something?" I asked. "Yes," was the reply, "I have lost my dog". "Well," said I, "Your dog is with me and it has been chasing my sheep. You had better come and make your peace with me." This owner was entirely different and full of apology and wanting to pay me compensation. I replied that there was no need as no damage had been done. But I did ask him to please learn something from the experience.

CHAPTER 26

BSE PART ONE

BOVINE SPONGIFORM ENCEPHALOPATHY. This was allegedly the new cattle plague which ruined our export trade and cost us billions of pounds.

Hitherto we had a condition in cattle known as the 'staggers', a very descriptive term which covered a multitude of conditions. When working at Gordon Road I would come in on a Monday morning and maybe there would be as many as ten cattle carcases and offals together with a number of sheep. These animals had been killed over the previous Saturday afternoon and Sunday. There were always one or two slaughter men on call over the weekend to deal with any emergencies. Any one of these cattle with the so called 'staggers' could have been harbouring BSE, who knows?

It was not until I was working at the new abattoir at Nailsea that this "new" cattle disease was officially diagnosed and declared and new stringent measures were brought in to be enforced.

We were told that it was caused by a damaged protein which like an anthrax spore was virtually indestructible. We were told that if people were to eat certain parts of an animal suffering from BSE they would certainly get Variant Creutzfeldt–Jakob disease. Now there has always been Creutzfeldt-Jakob disease (CJD) mostly affecting old people and another form of the disease in the South Sea Islands called Kuru disease, where apparently as a mark of respect the tribes eat the dead bodies of the deceased. The men allegedly get the best meat, whereas the women have the bones and skull where residually much of the main nervous system including the brain is to be found. The disease seemed to be spread in this way, with the women becoming infected more than the men.

There is a disease of sheep and deer called Scrapie we have been aware of ever since there were trees. Inevitably we have eaten sheep which have this disease, the visual characteristics of which have not yet manifested themselves. As far as I know there have never been any repercussions from eating animals with the sheep encephalopathy.

Now we were told by Professor Lacey among others that we were all going to die and every blade of grass in this country was infected with BSE. Interestingly I have not heard much from Professor Lacey in recent years.

The knee jerk reaction to all this by the politicians was to decree that all bovine animals which had more than two adult incisors had to be culled, even if the third tooth was only just visible through the gum. Like our children, they do not replace their baby teeth to some sort of order. It is generally accepted that the first two adult teeth appear at around 18 months of age but it is not inconceivable that that might be 16 or 17 months or just maybe 19 or 20 months. Thereafter another pair of adult teeth appear at six monthly intervals. As can be seen this is a very inexact science.

Sadly I had to reject one such animal because of a third tooth just appearing through the gum. Apart from the financial loss to the farmer who had obviously not checked carefully enough, before he brought the animal in for slaughter, this was an incredible waste of life. In my mind it is comparable to the millions of lives wasted on the cattle cull or the European dictat about bicatch in the fishing industry, dumping thousands of dead perfectly edible fish back into the sea, simply because the fishermen have exceeded their European quota. Bear in mind that Prime Minister Heath gave away our fishing industry to get us into Europe in the first place. We now have no jurisdiction over our own home waters.

All these dentition tests were done before passports were inflicted upon farmers to identify individual cattle from birth. Of course when so much hangs on under or over age at the point of slaughter it is critical for the birth date to be equally accurate. This entirely depends on whether the farmer has a good or bad memory, or possibly if he leaves dating the passport for a few weeks to give him a bit more leeway at the ultimate end. Farmers have to fill out so many forms today; they could probably solve the unemployment problem by employing full time clerks. This of course is the same for most of us in our day to day employment, meat inspection to name but one, it does not matter anymore how good or bad an inspector you are as long as you have filled out the right form. Why don't the people with shiny backsides let us get on with what we know best or will it affect their mostly non productive jobs?

Sheep of course do not yet have passports and the two incisor principle is retained. It is thought that they raise their first 2 adult teeth at around 12 months of age, but of course again this is an inexact science. In the case of sheep the slightest appearance of just one adult tooth is enough to make it an "overage". This requires a different selection of Specified Risk Material (SRM). Namely, the head, the spinal cord which means the animal has to be split down the middle, the spleen and the ileum (part of the small intestine). Whereas underage sheep have just 2 Specified Risk Materials, namely the spleen and the ileum.

The Meat Hygiene Service felt it necessary to appoint numbers of technical officers, to peer carefully at the mouth of each sheep, but latterly they decided that this was not that important and got rid of them again. It has to be said that throughout this debacle there has not been one sheep other than animals deliberately infected in the laboratory, which has been diagnosed with BSE, unlike the other horrendous disease Bovine Tuberculosis which although very rare I found in a sheep in 2010 which came back positive.

The SRM in cattle in the beginning was the head (the tongue could be harvested), the thymus glands in the neck and above the heart, the spleen, the intestines and the spinal cord. The spinal cord in particular caused me some amusement; the spinal cord extends from the brain to the tip of the tail. Unfortunately, the clever people who dream up what is going to be SRM initially forgot about the tail, and tails were being sold to the customers just the same as they ever were. There was obviously consternation at high level, and cleverly they came back with calling the last part of the spinal cord – chordae equina – which is of course the correct terminology. The cord splits into three when it arrives at the pelvis, and is described as the horse's tail and operates the back legs and tail. What a clever bit of face saving!

It seemed to me that tissue of very little value was selected to be SRM. Consider for a moment if this is a disease of the nervous system one would ask the question if the spinal cord is the trunk of the tree what about all the branches? Maybe if the disease is so bad then perhaps instead of playing with bits and pieces, we should have thrown the whole thing away. But of course the politicians could not be seen to destroy a whole industry, so they played politics and told the panicking population that all the dangerous material was being removed at the abattoir. The Agriculture minister at the time John Selwyn Gummer, was to be seen feeding his young daughter a beef burger, to prove to the people how safe it was. As time went on they played around with what was safe and what was not. Spleens were considered safe, so were the thymus glands. The spine itself became dangerous so you could no longer buy a T-bone steak but you could now make a stew from cattle cheeks. Only part of the tongue was now safe, but of course it all depends on the angle of the knife when removing the toxic waste part.

208 IT'S IN THE BLOOD

It sounds as though I treat what is a very serious subject in a very flippant manner. Make no mistake I feel as much as anyone for the people and families often involving a young family member who develops variant CJD. The tragic end is inevitable and very distressing for the victim and all the friends and family involved.

Controversially I do not believe that cattle, who themselves suffer greatly with very distressing symptoms, are to blame. I will enlarge on my theory in the next chapter.

CHAPTER 27

BSE PART TWO

WHEN THE DISEASE was first declared, I kept an open mind on the knee jerk reaction to the precautionary measures declared at the time, and more since. I am a very practical man, and inevitably suspicious of scientific pronouncements, which I have noticed over the years rarely stand up to close scrutiny in the real world. Predictions of the spread of foot and mouth by the scientists was a case in point.

Numerically it has to be said that the number of people who have actually died is extremely small namely 176. There are probably as many people killed on the roads in a week, as have died however tragically in the individual cases. The last to die being Holly Mills at the tender age of 26.

Holly was the last known sufferer of variant CJD and with her tragic death in 2011 after 8 long years of degenerative illness hopefully this has brought closure to this animal/human disease. The first known sufferers were in 1995 and the numbers peaked at 28 in the year 2000. There have been no known new cases since 2010. The scientific experts predicted a final death toll at somewhere between 200 and 3.5 million. Thankfully the disease claimed less than 200 but again

I would say where does the blame for those tragic deaths lie? With the hasty withdrawal of the wide use of organo-phosphates, will we ever know? I doubt it.

It has been said that feeding dead cows back to other cows was behind the outbreak of disease. It has also to be said that the sterilising process of this material has changed dramatically from that which I knew when I started my career. I described in an earlier chapter how the waste material was exposed to high temperature for a very long period in order to extract the tallow(fat). The cooked flesh and bone was then ground up and incorporated in cattle feed. All the small companies at the time were bought out by a huge German company who exposed the same material on a conveyor belt system to high temperature for a much shorter period. Inevitably this cannot be such a good sterilising process, even though it was agreed to by the politicians of the day. However, it seems unlikely to be the cause of the dramatic spread of BSE given the relatively small amount of possibly infected material in amongst hundreds of tonnes of waste tissue going through the rendering process.

Now, assuming that the infection came through this route, surely all cows fed on this material should have become infected, but of course they did not. Only a very few seemed to come from each affected farm. I actually found a BSE heifer which was brought into Nailsea, and the farmer who was known to me released this animal into the lairage. I was talking to the owner looking over his shoulder at the animal beyond. I noticed this animal was looking in our direction but was not focussing on us, but somewhere in the middle distance between us. She was also very agitated and was continually licking her nostrils. I said to the farmer that I was of the opinion that he had a problem. he asked what I meant and I said "I think you have got a case of BSE". He said "We have never had one on the farm before"; I said "Well I think you have one now".

At the time, the O.V.S was only part-time, and this particular one worked for the rest of his time at Langford veterinary college. The procedure then was for the vet to look at the animals which had arrived in the morning for slaughter, but at that time I was allowed to look at anything which came in later in the day.

I called the vet at Langford which is what my instruction was if I did not like the look of anything. He came over and agreed with me about my possible diagnosis. A Ministry vet was then called and he decided to dispatch the animal then and there with barbiturate injection, to be collected later for analysis. An identification metal tag was put through its groin skin.

Interestingly the Langford vet claimed that he had found the animal. I drew him to one side and asked him how he could have found it as he was 7 miles away at the time. If I had not made my diagnosis that animal would have gone into the food chain. One wonders how many more have already done so. Later the results from this animal came back as positive. I will return later in this chapter with interesting information concerning the owner of this animal.

Although strict regulations were in place as to what could and what couldn't be allowed out from the abattoir to be sold, in the beginning, all specified risk materials (S.R.M) had to be weighed, stained 'blue' as against 'black' for rejected material and placed in separate containers for later collection, or, as in the case of Nailsea where an incinerator was installed to burn all the S.R.M on site. It was insisted on by the Ministry that all Nailsea S.R.M had to be stained "blue" after being placed in the incinerator. Begging the question why? Subsequently the resultant ashes had to be buried in ground never to be grazed again by cattle or sheep.

All this to satisfy the pen pushers, but somewhat amusing when you see all this material so carefully separated and then when collected being tipped into the same bulk lorry with all the other waste material. In fact, when once working at the big cull abattoir on the North of Bristol fringe, I was asked to supervise the separate boning area on the site which carried on working on meat for human consumption, while the cull continued on another area of the site. Sides of beef were imported from Europe and Southern Ireland. Because the S.R.M rules in these areas were different to those in this country, all this imported meat had to be boned out under Meat Hygiene Service supervision and all the bones stained 'blue', before being removed from the site. They would then be tipped into a bulk lorry together with all the other non-SRM bones from cattle slaughtered in this country – you could not make it up!

Another interesting aside was that the health marks from the country of origin were all removed so that the resultant beef could be sold as if produced in this country – so much for traceability!

While still working for the council at Nailsea I received a phone call one day from the Environmental Health Office enquiring as to how large were the seepage holes in the drain traps in the abattoir. These traps are stainless steel buckets full of holes placed in each drain to collect pieces of tissue to prevent them from going into the sewage system. At intervals these traps would be removed and tipped into the S.R.M containers and the general fluids would pass on down the system. I measured the holes and found then to be half an inch in diameter. As a matter of interest, when I returned this information to the office on the phone I enquired as to why they wished to know. I was told by the girl on the other end of the phone that they had to be sure that no S.R.M was escaping into the sewage system. I asked her how big she thought was a grain of sawdust produced when sawing down a bovine animal, which would inevitably contain spinal cord material, and while she was thinking about that, how big did she think was the prion (faulty protein) which causes BSE. I explained that you could get a billion grains of sawdust all at one time through one hole in the bucket and there were probably 50 holes in each bucket and five separate drains. She went very quiet and needless to say I heard no more about the size of the holes in the drain buckets. Multiply that by the number of abattoirs and the number of drains all discharging into the sewage system up and down the country – you could not make it up!!

Consider how much atomised spinal cord is discharged into the atmosphere in just one day in abattoirs up and down Britain, some plants doing 500 cattle a day or so. This is caused by big 'chine' saws, water cooled, which cut a beast in half in just seconds. The water coolant atomises and takes some of the SRM into the atmosphere in a breathable vapour.

It is obvious therefore that abattoir employees are expendable when it comes to risk. Perhaps the clever scientific people hadn't thought of this one. But it is interesting that not one person working in the abattoir environment to my knowledge has developed variant CJD. While engaged on the cull scheme as I was, where many thousands of old cows were murdered because of their age, the risk (if risk there was) was far greater because of the age of the animals involved.

Eventually during the scheme someone somewhere thought that perhaps it might be a good idea to saw these cattle down to one or other side of the spine instead of down the middle involving the spinal cord. Well done!! But that does not stop all cattle for human consumption being sawn down the middle.

If, as is suggested that all those inhaling or ingesting BSE infected material will develop Variant CJD, then surely it must be about time for some of us old men to start falling off the perch as predicted by the science. We are also told that it could take years, but if that is case how much longer must I wait? I have only been doing the job for 50 years. During the whole period of stringent regulation I have continued to eat cattle thymus glands, a particular favourite of mine, just to see what would happen, nothing has.

The answer must be sought elsewhere. There was a parasitic fly which used to lay its eggs on the hair of the legs of cattle around the early summer of the year. Cattle instinctively knew when these flies were about and would run around the fields with their tails in the air. There must, I think, have been something in the frequency of the drone that the flies made, when near cattle. The old farmers (particularly here in Somerset) would say "Ah! The fly be about".

These flies, called Warble flies, (hyperderma bovis and hyperderma liniatum), would lay their eggs and the grubs would hatch out and bore their way through the skin. They would then travel under the skin right up to the throat, and live for a period of their development between the red and white muscle of the oesophagus, (food pipe). After the required period they would migrate again to the hollow of the animal's back and start to grow into large maggots about the size of the top joint of a small ladies' little finger. They would bore another hole through the skin to breathe through and live for a period in a capsule of pus under the skin.

In a bad infection one might see as many as 20 of these maggots on one animal. The condition was known colloquially as the 'lick' as cattle would be so distressed by the irritation caused that they would try to strain their heads back to 'lick' the affected area to ease the itching, covering their flanks with saliva. They would also try to find an overhanging branch under which they could scratch their backs, sometimes so violently that they would bruise the whole sirloin area of the back. When I was a young lad on farms I found it was possible when cows were chained up for milking, to squeeze the lumps on their backs and the maggots used to shoot out on a jet of pus through the breather holes. This procedure used to cause the cow some distress, and she would arch her back accordingly. Later of course, she would have felt relief from the irritation. Normally, in the fullness of time the maggots having completed that stage of their development would crawl out through the breather hole and bury themselves in the ground to pupate. The next generation of flies would appear in due course and the whole cycle would begin again.

I have taken my reader down this long route to explain in detail what was to happen next. The damage this parasite caused to the meat industry was in the form of bad bruising, to the sirloins of cattle. The damage to the hides by the breather holes ruined them for the leather industry, also there was the extreme distress caused to the cattle which I am sure affected their productivity.

The Government of the day actually came up with a good idea. We will get rid of the Warble fly. A very commendable decision, but at what wider cost?

This was in the 1970's and 1980's; geographically there was a higher incidence in the southern half of the country around 40% and only about 20% in Scotland. The eradication began with the Warble fly (England and Wales) Order 1978 which required compulsory treatment of infected animals in the spring, and recommending voluntary treatment of susceptible cattle in the autumn. On 15 March 1982 the Warble fly (England and Wales) Order made Warble fly infestation in cattle a 'notifiable' disease. This gave M.A.F.F (Ministry of Agriculture, Fisheries and Food) veterinary inspectors power to serve compulsory treatment notices on farmers whose herds were found to be infected by Warble fly. It also allowed inspectors to restrict the movement of cattle until treatment had been completed. In 1993 the Warble fly returned in imported cattle and those found to be infected were returned to their country of origin. (Under the 1978 Order cattle were required to be treated with Phosmet, Famphur or Penthion all of which were (pour on) organo-phosphates or Derris powder, a contact insecticide.)

There followed a further Order in 1983, the Warble fly (England and Wales) (Infected areas) Order 1983. This allowed for areas still found to be infected to be declared infected areas and in 1984 six such areas were all declared in the South of the country and in Wales. Further Orders in 1985 and 1986 came when three more areas all in the South West were declared. By 1985 the figures were reduced to 0.01% from a figure of 38% in 1978.

When the Warble fly eradication scheme began and the disgusting little parasite became enemy number one, they also became a 'notifiable' parasite. There are several 'notifiable' diseases which if suspected must be declared to the Agricultural Ministry, Foot and Mouth, TB and Anthrax to name but three. The method of Warble fly extermination was to pour on organo-phosphate all over the backs of cattle to kill the maggots. Organo-phosphates are very good at killing small parasites; take the case of aerosol sprays to kill flies in the house. The flies go into kamikaze mode with their nervous systems destroyed. They crash into walls before they die. We of course are breathing in this poison.

Then of course there were the mosquitoes and small invertebrates that were annoying our troops in two Gulf wars. Could Gulf war syndrome be the result when the troops were sprayed with organo-phosphates?

Farmers used to have to dip their sheep by law for sheep scab (another small parasite) and to deter the predations of the green bottle fly which lays its eggs on sheep resulting in maggot infestation which, if left, will allow the unfortunate animal to be eaten alive. Each sheep was immersed in a tank of organo-phosphate. Each sheep had one immersion. The farmer or shepherd had as many doses as sheep dipped when the sheep shook themselves when they emerged from the dip tank.

There used to be a product dissolved in water to dip your dog in for flea infestation. You would not see another flea all summer. You can no longer buy this product. It is no longer a legal requirement to dip your sheep and if you decide to dip them anyway the dip used is not organo-phosphate based. Warbles are no longer a problem and the eradication scheme worked very well. It is no longer a legal requirement to treat your cattle. What about Gulf war syndrome the existence of which, was vehemently denied by the Government of the day? These poor men were left to die miserably with various nervous disorders. The last few left alive were given a few thousand pounds to go away.

There was a farmer down in Somerset by the name of Mark Purdey who had a herd of Jersey cattle. He flatly refused to let his cattle be soaked in organo-phosphate. He also had the temerity to argue his case with the Government's official dictat. One of my biggest regrets was not to go and meet this man, he did a lot of research into the use of organo-phosphates in the environment and he could see the possible dangers. He even travelled to Canada where there had been an explosion in a organo-phosphate factory, where some deer living in the vicinity developed symptoms not unlike BSE. Very sadly he died several years ago before I thought to go and meet him and swap theories.

I do know that a slaughterman working at Weston-super-Mare abattoir by the name of Arthur Durbin had a predisposition to react to the particular organo-phosphate used in sheep dip. We would have batches of lambs come in, which had been dipped in the not too far distant past; you could tell this by the smell on their wool. They were not meant to be killed for some time after dipping but of course if the price is right in the market, the farmers will avail themselves of the opportunity. If Arthur so much as stood in front of or touched one of these fleeces his lips would go sticky and his arms and face would swell up and he would have to go home until the problems subsided. He was progressively ill for a number of years and in the end it contributed towards his death. Of course no-one would commit to a connection with organo-phosphate, but we had watched him degenerate over quite a long period.

Returning to my farmer friend whose heifer had been diagnosed with BSE at Nailsea, he was himself taken seriously ill after dipping his sheep. The problem has debilitated him terribly in that he suffers from a condition known as dilated cardio-myopathy. The medical fraternity who treated my friend for his condition have said that there were strong indications that exposure to organo-phosphates can cause this sort of damage to his heart, but it would be difficult to prove as little of the chemicals would remain in his body after 24 hours had elapsed. My friend told me that he was one of ten children from a strong farming background, including him three of the boys developed dilated cardio-myopathy which went on to kill one of them and severely debilitate two of them. All three of them had been exposed to organo-phosphates during their farming lives, when dipping sheep.

It could have been a genetic pre-disposition to cardio–myopathy or a disposition to react to organophosphates. But, allowing for a predisposition to organo-phosphate reaction would seem to me to fit the facts. Only some cows react to organo-phosphate contact and only a relatively few people react as I have described. I do remember at Gordon Road when a lorry came in loaded with sheep that had just been dipped but the concentration had been wrong. I have seen this before when some farmers are given instruction as to the dosage of all sorts of chemicals used on the farm, (and that in itself is a controversial subject) some will work on the principle that you cannot have too much of a good thing and up the measured dose on all sorts of things.

The sheep in the lorry were falling about as if drunk; several had already died. I called the council vet to decide what to do with them; he did not really want the responsibility for around 50 sheep, so he drove away saying you must do whatever you think. I decided to have all the sheep slaughtered and put in a skip. In doing so I may have put these men's lives in danger as the sheep were wringing wet with dip.

The problem then arose as to why young people seem to be predominantly affected by this variant CJD, bearing in mind that CJD was known as a disease of old people and took years to develop Suddenly I realised that with a dramatic upsurge in children with head lice at school during the 1970's organo-phosphate shampoo had been used to eradicate the parasite. Acknowledging the aforementioned predisposition severe reaction is experienced by some individuals. In fact the son of another farmer I know having had organo-phosphate shampoo used on him for the removal of lice and nits was taken very ill and had to be rushed to hospital to save his life.

Of course all my theories will not appeal to the political class. During the bad time of BSE thousands if not millions of animals were slaughtered and burnt. The meat and farming industry was devastated by bad publicity, people gave up eating beef, some of whom still do not eat beef to this day; of course they should have given up eating sheep as well but this connection was never made by the general public.

The ongoing cost of policing the specified risk materials in the abattoirs and beyond must be huge. The cost of blue dye alone must have been exorbitant. If only I had had a bit of preknowledge of what was to come I could have bought some shares in the blue dye company. The cost of disposal must have run into billions and billions of tax payers money; we were all told by the politicians and their advisors that all this devastation was necessary to protect us from certain death.

The fact that using organo-phosphates to get rid of the little Warble fly might just possibly have something to do with the problem, was never mentioned in public. Consider where the politicians end up when they finish running our lives, which can be on the board of directors of huge chemical companies who amongst other things make the dangerous organo-phosphates. Imagine the compensation due if my theory could be proved. But as they love to say we will draw a line under it and move on. Of course all or almost all of the evidence has long since been burnt. Does it matter now that they destroyed millions of cattle lives for what? And the damage to the meat industry and beyond is incalculable.

All this secrecy and misinformation is typical of how the politicians work whenever something controversial comes along. Bad news is usually buried under something less controversial. This puts me in mind of a painter and decorator who was refurbishing Torbay hospital. The job was nearly complete when one of the resident surgeons came down to see how things were going. He said to the painter and decorator, "I suppose a bit of putty and a bit of paint will cover up a multitude of sins in your game". "Yes "said the painter, "And I expect six feet of soil covers up a multitude in yours".

CHAPTER 28

MYCOBACTERIUM TUBERCULOSIS AKA TUBERCULAR BACILLUS OR "TB"

THERE ARE KNOWN to be four types of this particular bacterium. There is of course the human form which years ago was known as "consumption", and killed thousands of people. This was largely brought under control with the advent of antibiotics, but now has cleverly mutated so that at least one strain is antibiotic resistant which allegedly originated in the Russian gulags. In recent years the number of cases of human TB in this country has been increasing again due to mass immigration and the global movement of people.

The bovine form of this disease, originally known as "pearl disease", was recognised as far back as early in the 19th century and was clearly an endemic problem even then. The progression of this chronic disease normally follows one or more routes into the body visible on post-mortem examination. It can initially affect the head manifesting its presence in the lymph glands in the form of a yellowish caseous sometimes purulent lesion. It can then progress into the lymphatic system of the lungs as well as the substance of the lungs. Another route is via the digestive system, in particular the lymphatic glands draining the intestinal area, often also involving the liver. From these

MYCOBACTERIUM TUBERCULOSIS 221

early beginnings it can move throughout the body ultimately killing the animal concerned.

The third form of the disease, described as "avian TB" primarily affects birds, but at one time was commonly found in pigs where they were run with poultry. There have been a few recorded cases of humans suffering from avian TB but this is very rare.

A fourth type of TB occurs in fish, particularly the halibut, and in other cold blooded animals, but is thought to be harmless to warm blooded animals and therefore man.

To return to bovine TB which we are primarily concerned with here, it is readily transmissible to man via the consumption of unpasteurised infected milk or undercooked infected meat. TB was probably the primary reason for meat inspection coming into existence. This was initially carried out by the local council surveyor who at that time fulfilled many functions as well as building control. As time moved on sanitary inspectors came into being who were then followed by public health inspectors and today we have Environmental Health Officers. All the above performed meat inspection as part of their duties and it was not until sometime later that courses were run for specialist meat inspectors such as myself.

At the current time, the procedure for testing cattle is that a vet from DEFRA or a private practice visits farms and tests all cattle at least annually. The test is performed thus: an injection is put under the skin, this is then left. After a period of four days the vet returns and sees if the animal has reacted to the injection. If it has the animal is a 'TB reactor' and must be culled. If a herd has a TB reactor in it, there is a fear that other cattle may be infected but not have reacted to the test. Consequently the whole herd is put under a movement restriction and has to be kept on the farm. The test is then repeated after 60 days. Any animal found to react is sent for slaughter and the 60 day restriction is repeated. If no further reactors are found the herd is maintained under a movement restriction for a further 60 days when the test is repeated. If this second test is clear the herd is declared 'free of TB' and the movement restriction is lifted.

* * *

Many years ago when I was very young in the fifties bovine tuberculosis was an endemic problem in this country. It was decided by the Government of the day to attempt to eradicate it by testing all cattle and slaughtering all the resultant reactors. This was a bold plan and was set in motion county by county and as far as I can remember it started in Pembrokeshire which coincidentally is my home county. When I eventually started my career as a meat inspector I learnt from older colleagues that this action resulted in a considerable reduction in the number of cases found in abattoirs.

The improvement remained for a number of years, but then there appeared to be outbreaks in certain areas, firstly in an arc around Bristol starting in Gloucestershire, out through Bath and back around to the Failand ridge. This latter is the hill behind the M5 where it goes up on stilts in the Gordano Valley. Other outbreaks occurred in Cornwall and mid to southwest Wales. Investigative work must have been done resulting in the conclusion that the bovine strains of TB was being harboured in and spread by local badger populations.

Any badger setts which were local to an outbreak would be gassed and teams of men were put in place to do this work. One particular sett near Dursley in Gloucestershire for some reason attracted a large amount of publicity as people became aware that badgers were being killed to protect cattle. Dead badgers in the vicinity, probably road traffic accidents or possibly individuals which have succumbed to the disease and stool samples from the local badger toilets tested positive and the decision was taken to gas the sett. Protesters appeared with placards but the job was completed and to see justice done a JCB dug out 16 dead badgers which in due course were tested. Unfortunately not one was found to be positive. This incident gave ammunition to the badger protectionists who would wish that not a single badger would be killed with no regard whatever for how many cattle have to die. From that very day subsequent Governments have sat on their hands and done nothing and the disease has proliferated to epidemic proportions.

* * *

It seems that the general public are either oblivious to what is going on, or they just do not care that thousands of cattle are being slaughtered before their time and that many farmers' livelihoods are being put at risk because of successive Governments' failure to address the problem. Thousands of dairy farms have already gone out of business over the last few years as the price they receive for their milk is governed entirely by the stranglehold the supermarkets exert with their massive buying power. It seems the only way some dairy farms will survive is to hugely increase the number of cows they milk, turning a very small profit per litre, but on a vast scale. Unfortunately this can have repercussions for the environment. Previously the national herd was spread thinly over vast areas of the country, but the trend towards these much bigger units housing hundreds if not thousands of cows concentrates potential pollutants ranging from the delivery of huge numbers of tonnes of the feed necessary to the problems of disposing of vast quantities of slurry. Also, from an ethical point of view, is it right that many of these animals will never see a blade of growing grass?

From a disease point of view, infections such as TB, spread rapidly when animals are in close confinement, therefore these large units are at particular risk. It seems that the further we move away from nature, the more problems we incur.

A friend of mine has a semi-intensive modern dairy unit in which his cows are housed during the winter. Unfortunately he had 75 reactors in two consecutive TB tests. It was thought that a badger or badgers were gaining access to the cattle shed through a gap under the large doors. The cattle were being fed on stock feed potatoes and maize silage which was attracting the badgers and they spread their contaminated saliva and urine up and down the food alleyway.

Denying the badgers access to the shed alleviated the problem for some years but sadly since then he has had several sporadic outbreaks on an adjoining farm, involving young cattle some of whom had never been out, but possibly badgers could have gained access to the building where they were housed.

At the time of his first outbreaks I was working allegedly to control BSE on the 'over 30 months' cull scheme whereby no bovine animal over 30 months old went into the food chain but instead was slaughtered and burnt. This of course included tubercular reactors which were all killed on one particular day per week. One of his cows which passed the first test and was not even an 'inconclusive' (an animal which is only suspected of maybe having TB) failed the second test. On examination at the abattoir the animal was rejected as being unfit to cull under the OTMS, due to her generalised condition of TB (this meant the farmer didn't get OTMS compensation but still lost the animal). In my opinion there was no way this animal could have developed such symptoms in the time between the two tests, which calls into question the accuracy of the test itself.

* * *

For a number of years dead badgers were collected and tested by the Ministry of Agriculture Fisheries and Food for evidence of bovine TB. These were mostly road traffic accidents because by now badgers had become a protected species. I became involved in the collection of many badgers with the assistance of my shooting friends around the Southwest of England. Each body would have a brown label attached to its neck stating date and where found. We did this on the understanding that we would be informed about each numbered animal as to whether it was positive or not. The testing took about 6 weeks. Sometimes I might have collected up to ten badgers at a time. A friend in the Ministry at the time, when collecting some badgers from me, pointed out what he described as a 'driven' badger. This was an individual where the hair had been scratched from the buttocks and rear end. Other individuals may exhibit fresh scratch marks or scabs in the same areas. I then realised I had seen 'driven' badgers before and subsequently made a mental note that many if not all of these tested positive. My conclusion from these observations was that badgers, much like pigs, may have the ability to sense and drive away a diseased animal. Badgers are of the musterlid or weasel family and have no connection with pigs, despite being commonly known as 'earth pigs'. It is generally known that pigs will attempt to drive away a sickly individual, and if confined in a building or pen may well kill the unfortunate animal – this, we presume, is an instinctive defence by the group against the spread of disease – a natural cull if you like.

Badgers are very territorial animals so an individual who may have been driven out of his home setts for reasons of illness could only go to the periphery of the territory of those setts, otherwise he would be straying onto the next door range. He would therefore probably have to live alone at the edge of his home range. I spent some time trying to make sense of the incidence of bovine TB and driven badgers. Sometime later I was invited to listen to a lecture given by an eminent scientist who had been looking at the connection between badgers and bovine TB in cattle. Experiments had been done in Cornwall, a hotspot for the disease, whereby adjoining badger ranges were differentiated by the feeding of different colour small glass beads in peanut butter, an apparent badger delicacy. The beads would manifest themselves in badger faeces in the various badger toilets contained in the home range so the scientists would know who lived where. A TB breakdown in cattle would occur in let's say red beads territory. The response would be to gas red beads setts, and for a while thereafter apparently things would quieten down in that area. There were then further cattle outbreaks around the periphery of red beads territory, and after a further time lapse bovine disease started to recur where it first started.

As he explained this I realised that what he was saying fitted my theory about the driven badgers. An infected badger or badgers, having contaminated the centre of the home range could then have been driven to live in isolation on the periphery of the territory where they then infected that area. After the sett was gassed, and if they lived long enough, they could return home as there was nobody there to stop them, and would proceed to re-infect where they first started.

Only four questions were taken at the end of the lecture and mine was lucky enough to be one of them. I explained my interest in the subject both as a meat inspector and as an erstwhile collector of many badgers for testing by the Ministry in Gloucester. I said that I had noticed a number of driven badgers amongst the ones I had collected and that many, if not all came back as positive for bovine TB. I asked him if any work had been done on driven badgers and he said it had not, to which I responded by respectfully suggesting that he might.

The implication of all this is that if the Government had acted decisively decades ago it might only have been necessary to kill the so called driven badgers and not necessarily to annihilate all badgers in a given area. This could be achieved by cage trapping and selective killing, as long as the cages were not tampered with by the "don't kill any badger" brigade, or alternatively by night shooting any badger which showed any signs of coat damage around the rear area.

Very sadly, all the good work that was done back in the fifties eradication scheme has been wasted. This is because when there were outbreaks in three well defined areas of the country insufficient action was taken to as far as possible eradicate badgers in those areas, even though scientific evidence pointed towards badgers being responsible to a large degree for the spread of the disease. This was because the politicians of the day were terrified of losing votes if the public at large came down on the side of the anti-killing badger brigade whereas the farming community are numerically of little consequence. So instead the Ministry opted for delaying tactics by all sorts of crazy schemes like doing nothing in some places, killing some badgers in some places, and allegedly trying to annihilate the badgers in other areas. However interesting these scientific trials may have been, it allowed time for this insidious disease to proliferate. Moreover, a farmer, some of whose cattle have to be slaughtered and who is subject to rigorous movement restrictions, on discovering he is in an area where no action is being taken against what he perceives as the enemy, the badger, will be tempted to take his rifle or shotgun and drive around his farm at night and shoot every badger he sees, and then place them on the road and run over them a few times. This action, if undertaken, clearly negates the validity of the trials. I brought this particular point to the attention of the Professor giving the lecture and asked him if he was aware that some farmers were taking the law into their own hands in this manner and he said he was fully aware of the situation.

The collection of dead badgers by the Ministry has now ceased, allegedly for Health and Safety reasons and to save money, but of course as everyone knows if you are not looking for TB in badgers you will not find it. It was possibly done to appease the "save the badgers" lobby, who at the time had groups of people picking up dead badgers on the road to stop them being collected for Ministry testing.

* * *

There has been much said about deer and the spread of bovine TB. My first personal experience of this came one evening many years ago when a game dealer and stalker friend of mine brought me the carcase of a roe deer he had just shot on the Mendip hills. Knowing what I do for a living he wished me to examine it. Sadly he had left the offals out on the hill but on examining the internal surfaces of the thoracic and abdominal cavities I saw what I believed to be tubercular lesions adhering to both surfaces. These are colloquially known as tubercular 'grapes' because habitually they hang in clusters. I have seen this phenomenon many times in cattle, usually when the animals are in extremis, but this was the first time I had observed it in deer. My advice to my friend was that he should inform the Ministry of our findings because suspicion of bovine TB is notifiable, and on no account should he attempt to sell the carcase. In due course my initial diagnosis was confirmed, and it was in fact the first case to be discovered in a roe deer certainly in the Bristol area. It had however already been found in both fallow and red deer elsewhere.

As I mentioned, my friend had already gralloched the deer in the field and buried the deer offal. For TB control, the practice of leaving deer offal where the animal was shot is not a good idea because of the obvious risk of spreading this serious disease. Animals such as badgers can smell the buried offals and dig them up, and even if they don't, at some point the land may be grazed by other deer or by cattle and thus spread the disease.

Since those days deer stalkers are advised to learn about, amongst other things, the meat inspection aspect of the animals that they shoot. They are given very basic training in a subject that takes years to be fully conversant with, which sadly may allow animals with serious disease to enter the food chain and may allow diseased offals to be buried in the field. Wouldn't it be better to put the offals in a bag and take them home?

After the discovery of TB in the roe deer I mentioned earlier, a particular veterinary friend of mine in the Ministry thought further investigation should take place. At the time large sums of money were being spent on buying and killing so called cohort cattle. These were animals which were closely related to BSE victims, and the hope was to arrest the spread of the disease if there was a genetic link. This left no money for the TB investigation, and I was asked to carry out, free of charge, a gross post mortem examination of heads and offals of roe deer which were killed within certain map references on Mendip. Black plastic bags would be deposited at the bottom of my drive labelled with date and where shot. Over a period I looked at several hundred of these offals and found a few with suspicious lesions, samples of which I would send to the Gloucester MAFF office. Very few came back as positive for bovine TB and still fewer came back as positive for avian TB. Still some were inconclusive for TB. The theory at the time was that these wild animals had stronger immune systems that were sometimes able to surround the bacillus and kill it.

At around the same time I was asked to give a talk by my game dealer friend to some of his fellow stalkers on the subject of meat inspection but slanting it in the direction of the evisceration and inspection of deer. To this end a fallow deer was shot earlier in the evening on a famous local estate in Gloucestershire on which I could demonstrate the necessary routine. Amazingly, because the odds were thousands to one against, I found tubercular lesions in the mesenteric (gut) lymph nodes of this animal.

Interestingly I also found lesions, particularly in the gut lymph nodes of the intermediate stage of a dog nasal worm called Linguitula Serrata. This is a parasite of which the Ministry seemed to be unaware until I sent them lesions for their perusal, and then they expressed surprise at the frequency with which it appeared to be occurring in deer. The method used by the laboratory to discover the presence of TB involves emulsifying lymph nodes from around the body and the resultant material is applied to agar plates where the TB bacillus will either grow if positive or not if negative. This procedure, although perfectly efficient, will not pick up an unusual parasite such as Linguitula, but over many years I have examined and thinly sliced millions of mesenteric or gut lymph nodes in cattle and have infrequently found lesions of the parasite.

I have also many times examined these same lymph nodes in rabbits and found it to be fairly common. Dogs and foxes can become infected by killing and consuming rabbits and harbour the adult worm in their nostrils. This worm can live in the dog or fox for about two years and remains in place by the use of hooks. The male worm is white in colour and about ¾ inch long whereas the female is darker and 3 or 4 inches long with a brown colouration in its centre due to the presence of numerous eggs. The eggs appear in the nasal secretions six months after infection and spread from the dog or fox by sneezing or nasal discharges. The number of eggs excreted can be up to half a million so clearly the pasture can become heavily infected. Herbivorous animals such as sheep, cattle, goats, hares and rabbits become infected by ingesting some of these eggs with the grass. Deer obviously become infected in the same way and it is clearly not advisable to leave viscera potentially infected with this parasite and/or TB on the hill for dogs or foxes to scavenge. Equally, the practice of leaving rabbit guts in the field is asking for trouble.

As recently as April 2012 I was asked by a man I had met only once before to go and put down an injured roe buck thought to have been hit by a car and subsequently pulled down by a dog. The man who sought my assistance with the dispatching of the deer said he would drive me to the scene in his vehicle. I was expecting to find the casualty by the side of the road or in the hedge or ditch adjacent but in fact he was 500 yards up a grass field lying on his side with a huge rip in his hide exposing much of his rib cage. I assumed this was as a result of the dog damage. I dispatched and bled the poor beast and we loaded it into a container in the back of the man's van. I enquired whether he would like some venison to which he replied that he was not fond of venison but had a friend who was. It is my habit when dealing with road traffic accidents involving deer, to try and save whatever I can for the person concerned. It is surprising sometimes how much can be salvaged and in others, particularly when the animal has been rolled under the car how little, usually when the stomachs have been burst and the contents forced around the body.

In this instance I was just about to go out for the day, and only had time to eviscerate the carcase before leaving, intending to continue to dress it on my return. Later in the day I started to skin the fairly large buck. Almost immediately I knew something was wrong, as I lifted it from the hanging position down onto a legging cradle I realised it was not heavy enough for its size. Closer examination inside revealed one of the lymph glands (internal iliac) which should have been about the size of a baked bean in an animal of his proportions, was in fact as big as a walnut. TB was my immediate thought and I returned to the green offal I had removed in the morning to examine in particular the mesenteric (gut) lymph glands. Careful slicing revealed three or four tiny (pinhead) sized lesions. These could easily have been missed without very careful incisions. Continuing my examination I found four pea sized lesions in the spleen and about ten in the lungs. The liver surprisingly was free of visible lesions but a further look at the outside of the carcase revealed a hitherto unnoticed huge lesion in the pre-scapular lymph node which drains the shoulders, which again should, when normal, be about the size of a baked bean but was in fact the size of a small apple and full of tubercular lesion and pus.

As suspected TB is one of several notifiable diseases and parasites I contacted a friend of mine in the Ministry so that he could collect some of the suspect material for verification. This process takes about six weeks and my friend said he would inform me as soon as the results were known.

Of the remainder I selected some of the tissue to be pickled for my collection of morbid specimens. Sadly I may not any longer take fresh suspected TB specimens to enhance my lectures at Langford Veterinary College. This is for Health and Safety reasons apparently. One wonders how young vets and other students can become familiar with tubercular material when they are only allowed to see photographs and pickled specimens. It must surely be an anomaly that inspectors such as myself sometimes have to handle this sort of material on a daily basis. I personally have probably seen more TB lesions than any living inspector by virtue of my many years in the business.

The only time any attempt was made by the Meat Hygiene Service to protect their staff against TB infection to my knowledge was while I was working on the over 30 months scheme (B.S.E cull) around 2003 or 2004 when on Wednesdays we would be culling around eighty to one hundred TB reactors. The then senior inspector always treated me with a certain amount of reverence, knowing that I had been a senior officer many years before but was now only employed as a casual inspector. His instruction from above had been that all inspectors when inspecting TB reactors should wear a face mask. This directive appeared at first to be a sensible precaution but when the masks were issued they were spring loaded and folded with an elastic band around the ears to hold them on to the face. All was well until a facial movement would cause the mask to close on one's nose and mouth like a rat trap cutting off your air supply. This necessitated readjustment of the mask often when you have just been handling tubercular material. Putting an infected chain mail safety glove near to one's nose and mouth seemed to be an unnecessary risk. I refused further use of the mask preferring to keep my hands and the infected material at arm's length and by frequent hand washing under very hot running water. The senior officer was not quite sure how to proceed with direct insubordination on my part to the normally unchallenged dictat of the M.H.S. My fellow inspectors some of whom were also casuals and therefore unimpressed by bureaucratic nonsense, also decided that I was right, and the shiny trousers of the M.H.S were wrong and refused to wear these stupid masks. The senior observed that stubborn old buggers were very difficult to manage.

TB – part 2

Recently the culling of badgers in parts of Gloucestershire and Somerset has been given the go-ahead by the politicians. This has caused major excitement amongst the "don't kill any badger" brigade. The idea is to remove 70% of badgers from given areas over four years, starting with a six week trial cull in these areas.

I think that history tells us that if a particular colony of badgers is living in an area with a high incidence of bovine TB reactors the best way to tackle the problem is to gas all the setts in that area. Badgers live in very close contact with other members of their colony and therefore infected individuals pose a very great risk to other badgers as well as the local cattle and deer populations. Surely it is a kinder approach to kill the whole colony than to pick off a guestimated percentage of the group? How can anyone know how many badgers make up 70% of the whole without knowing how many you started with? This could only be known by many, many hours of watching each suspect colony. Even if you achieved the imaginary 70% cull how do you know whether or not the other 30% are the ones you needed to kill? I did hear that the RSPCA have thrown their hat into the ring by suggesting that milk should be marked with its area of origin, and advising RSPCA members not to buy milk from Gloucestershire and Somerset. This is I suppose to punish the farmers in those areas for trying to protect their cattle from this horrendous disease which has been allowed by the politicians to proliferate to epidemic proportions.

Predictably the so-called badger cull has been postponed until June 2013, on the grounds that there are now apparently too many badgers in the pilot area. This rather backs up my earlier observations that nobody actually knows how many badgers are there in the first place, so how can it possibly be achievable to kill 70%?

The whole sorry political debacle will probably be played out all over again next June. All this on top of all the other delaying tactics acted out over the last decade or more, allowing this insidious disease to proliferate both in badgers and cattle. If ever politicians can forget about the numbers of city votes they may lose then they may seriously address this disgusting problem in a sensitive way, which in my and many other people's opinion is to gas all the suspect badger colonies. Inevitably this will necessitate the killing of many, many healthy badgers. The blame for this tragedy can be fairly laid at the door of all those who either for political expediency or for reasons of genuine concern (with which I sympathise) oppose the killing of huge numbers of our indigenous wildlife.

The RSPCA purport to stand up for the welfare of animals in general but cattle who can suffer terribly from the effects of TB obviously do not come into this category. They have still less concern for some struggling farmers who for the most part look after their animals very well and become very fond of their dependent beasts. I became very aware of this when involved in the foot and mouth debacle when farmers broke down in tears when we slaughtered some of their particular favourites.

Badgers who have attracted so much public concern also suffer terribly from this horrendous disease. You only have to look at the photographs kindly donated by Dr John Gallagher, a retired ministry veterinary officer who was deeply involved with TB in connection with both cattle and badgers in Gloucestershire and later on in Devon. He was strongly of the opinion that gassing a whole infected colony was the kindest way of dealing with the badger problems. Nobody wants to annihilate all of our badgers, only the infected ones but this simply isn't practicable. It is inevitable that many healthy badgers will have to die in order to destroy the diseased individuals.

It could be said if the "don't kill any badger" brigade had not raised so much emotive passion years ago, far fewer badgers would have to die now. If the areas found to be infected years ago had been as far as possible cleared of badgers, I think the problem would not have escalated as it has. This was entirely due to the pathetic politicians at the time, and since, who were frightened of losing urban votes and largely sat on their hands in the vain hope that the problem would go away on its own. Very sadly the reverse is true and the problem has escalated out of control.

Recently I went to a meeting about TB and badgers at Taunton livestock centre. The meeting was addressed by a farmer and great countryman called Brian Hill who had had TB on his own farm. He maintains that he can recognise infected badger setts in an area as against healthy setts. He has achieved a reputation in his apparent ability and travels many miles to advise farmers as to where problem setts are situated on their land. He is very strongly of the opinion that healthy setts and colonies should be left well alone. This I agree with in areas where there is no TB in cattle and so no TB control necessary.

So many tests and trials have been tried over recent years, with no success at all, however interesting it may have been to the scientists concerned. Can you blame a beleaguered farmer for taking the law into his own hands and killing every badger he sees on his farm, regardless as to whether it may be healthy or otherwise, when he sees all this legal stalling by the politicians and the "don't kill any badger" brigade.

Brian has come up with another possible way. When you see the total lack of success in every other direction, I think he should at least be given a chance to point out where badgers should be killed and where not.

You only have to see the distress on the face of Adam Henson, when seen on "Countryfile", who farms various rare breeds of cattle in Gloucestershire when every time he has a TB test it seems he loses some of his much loved herd.

I wonder if any of my readers has ever thought what happens to the resultant carcases from these reactor cattle. Well, I can tell you exactly what happens. They would be taken to particular abattoirs selected in each area of the country to accept TB reactors from their particular catchment areas. They would be slaughtered in the usual way; post mortem examination would then be undertaken by examining in detail a routine selection of lymph glands which drain various important areas of the body. Examine in detail means to slice each lymph gland very finely so that very tiny TB lesions may be seen. If no lesions are found, which is often the case, a selection of these glands is sent to a DEFRA laboratory for examination to see if the test has come to a correct conclusion. The animals that have visible lesions in their lymph glands are easily and readily seen which can then provide an immediate judgement as to whether or not the carcase should be rejected. The glands without visible lesions which had been removed are sent off to DEFRA for further examination in the laboratory and the growing on agar plates. These may in the fullness of time, (average six to eight weeks) reveal further confirmation of the presence of TB to justify the killing of the reactor. The results and the whereabouts in the body are not fed back to the inspectorial staff at the abattoir. Even if they were, the time lapse of six to eight weeks would be far too long, because the carcase and offals would probably have been eaten, or, if not, would be long since passed their use by date.

The law requires that in order for the whole carcase and offals to be rejected for human food, visible lesions must be verified in two separate sites. This may be the head and the lungs and be a respiratory spread or the head and the intestines and be a digestive spread. Just occasionally the lungs and intestines are involved with nothing visible in the head lymph glands. On occasion the lymph glands in the carcase are also involved. All these permutations allow for the rejection of the whole carcase. However the finding of visible lesions in just one of these sites only necessitates the rejecting of the offending organ, and the passing of the remaining organs and carcase for human consumption.

* * *

On one very interesting occasion, while working at Nailsea on relief, I found TB lesions in the head of an animal, which was not a reactor but one in the general kill. I went straight away to look at the intestinal mesenteric lymph glands, the intestines and stomachs being the next organs to be removed from the carcase. I found TB lesions in the mesenteric glands. I then returned to examine the lungs, and again found TB lesions. I said to the other inspector with whom I was working, that the whole carcase needed to be thrown in the bin. He said that perhaps he should detain the carcase for the perusal of the owner. I said that if he was not happy to just drop it in the bin then he should ring his senior officer who was working at Bridgwater at the time. He rang as I had suggested and the senior (another whom I had taught) asked what Dave had said. "Drop it in the bin" was the response; "Drop in the bin then" was his rejoinder.

The phone call had been overheard by an English OV working at Bridgwater at the time. He said to the senior "No, just drop the head and offals in the bin and pass the carcase". The senior was not happy with this decision and he thought it necessary to ring the top vet for the country at the time. He described what lesions had been found in this unfortunate animal. This vet said "Throw it in the bin, without a shadow of a doubt".

The abattoir owner, during the time of all this activity, was initially told that the whole carcase and offals had been rejected. He was then told that he could have the carcase and lose the offals. He was then told that both carcase and offals should be rejected. Understandably he asked what the hell was going on, and the sequence of events was described to him. He then came out with an observation that was music to my ears. He declared that in future if a tubercular lesion was to be found anywhere in a carcase or offal in an animal presented to his abattoir, he did not want the resultant meat for his customers. He said if the OV was prepared to allow the application of a health mark to the carcase, he would offer it back to the farmer concerned to put in his freezer, however if the farmer did not want it, he would personally put it in the bin and the farmer would not be paid. Why can't that sort of common sense be prevalent in the political class who make the rules and the crazy bureaucrats who advise in the first place and carry out the madness in the second place?

* * *

As the law stands, when TB is found in the general run of cattle coming through an abattoir, most carcases are passed as food under the criteria that I have described. These are animals which have probably been tested in the past and have either become infected after the test or maybe they were some of the inevitable false negatives when tested. This is bad enough in itself but consider the reactors which are killed in selected abattoirs up and down the country. Even allowing for false positives and false negatives the vast majority have TB. Very, very few are completely rejected and usually only those which exhibit visible lesions in the necessary two sites. All the rest showing lesser symptoms and in many cases no visible lesions at all are passed for human food. I find this absolutely incredible, when you consider all the pernickety rules and regulations in all other aspects of abattoir management. Just take the specified risk material which is still so closely monitored in cattle and sheep when BSE has virtually disappeared. TB is at epidemic levels and we the public, probably unaware, are eating tubercular flesh from not only cattle which present with lesions in the general abattoir but also the flesh of reactor cattle which are known to have TB.

How the legal criteria can be based on such inexact science I could never understand, when you consider that many cattle that have reacted to the tubercular test often show no visible lesions at all. Therefore cattle which present with lesions but are not reactors must have the disease pretty badly and I do not know of any man living who can declare with any degree of certainty how far the disease has progressed throughout that animal's body.

For these reasons, I entirely agree with the owner of Nailsea abattoir who instinctively does not want to sell tubercular meat to his customers, any more than those customers would want to eat tubercular meat. I do not think the risk is very great, because TB is easily killed by the application of heat, but when you consider the stringent application of rules and regulations in other areas of the meat industry it just does not make sense.

My conclusion can only be that the powers that be know that the financial cost of buying all those reactors and throwing them away would be absolutely astronomical. The Government claims a large proportion of the expenditure back from the abattoirs that contract to slaughter the reactors in order to sell them into the trade. Much of the flesh from the mostly older animals ends up in the fast food business in the form of burgers. This is obviously sufficient incentive to the powers that be to turn a blind eye to the fact that tubercular flesh is being sold to the general public.

A neighbouring farmer has had a sporadic problem with bovine TB for years. It can be a very disheartening experience when, having had a clear test or even two, sometime later more reactors are found and the farm is closed down again. He had a number of farm cats to control any rodent problem. Two of these cats became sickly and eventually died. He had a post mortem carried out and both were found to be suffering from bovine TB and he took the decision to kill all his remaining cats. He farms in an area where other cattle farmers have experienced similar problems, and in which I found the second case in roe deer that I have seen.

* * *

I have already mentioned my involvement with roe and fallow deer with regard to TB. When Nailsea ventured into killing farmed red deer, it was not long before I began to find lesions in some of these animals, coming from as far away as Cornwall. The disease was confirmed by the Gloucester MAFF office where the offending glands were sent for confirmation. Apparently with farmed deer it is not the practice to test the remaining deer on the farm for further reactors, as is done with cattle when a case is found in the abattoir. If several cases are found from a particular deer farm all the remaining animals on the farm are slaughtered because containing and testing deer is considered to be too stressful for that particular species.

Goats have been known to suffer from bovine TB but I had never heard of it being found in sheep. Two or three years ago while working on relief at Nailsea I found what appeared to be TB lesions in the lymph glands and substance of the lungs of a lamb, one of a bunch of about twenty originating from near Frome in Somerset. Thinking that although the lesions looked very much like TB I dismissed my original reaction because sheep do not get bovine TB do they? Not happy, I knew that later on that morning a friend of mine from Langford Veterinary College was coming to the abattoir to collect some morbid specimens which we had collected for use as teaching aids for the young veterinary students. I asked him if he would take the lungs from this lamb to a pathologist friend of mine working at Langford. He was the man who was asked by the government to look at all species of British mammals other than badgers to see whether there were any other reservoirs of bovine T.B. I amongst many others had collected hundreds of dead mammals from red deer to pigmy shrews for his perusal. I knew he was just the man to diagnose the problem with these lamb lungs. Later in the day he rang me and confirmed my earlier suspicions. He said that if I supplied him with the name and address of the farm from where the lamb had come, he would inform the Ministry and send them the offending lymph glands. The OV at Nailsea at the time was not happy with this arrangement and insisted on going to Langford to retrieve the lungs, so that they could be collected from the abattoir by the Ministry at Gloucester, as is the usual procedure with a suspect bovine animal.

* * *

I found it a very interesting phenomenon that when lecturing to various students at Langford Veterinary College in recent years, under no circumstances must I take any suspected or confirmed TB specimens into the post mortem room to show the students, for health and safety reasons. Yet only a few miles away hundreds if not thousands of TB reactor cattle are being slaughtered and inspected and then most of them are sold for human food with impunity. You could not make it up!

A while ago the body of an alpaca was taken into the post mortem room at Langford. It was found to be suffering from advanced bovine TB. Near panic ensued and Langford was nearly closed down as a result. Bear in mind this was a dead animal, taken into a post mortem situation for a final diagnosis. What else is a post mortem facility for? Alpacas and other members of the camel family are particularly dangerous from a human point of view because these animals tend to spit at those who upset them. Saliva apparently is one of the worst fluids to spread the bovine TB bacterium.

Obviously meat inspectors and slaughterman are expendable when working on tubercular reactor cattle and also when working generally on beef lines. Young veterinary students however must not under any circumstances be exposed to a bovine tubercular risk. But then we the general public are of such little consequence to politicians that they happily allow us to eat tubercular flesh. This is an example of the topsy turvy health and safety system in which we operate today.

* * *

Recently Brian May, famous guitarist in Queen, has enter the fray with I think the best of intentions. He observes that we should vaccinate our cattle instead of as he puts it the "cruel culling of badgers". Vaccination is out of the question, if we wish to maintain our disease free status in the world market. Exactly the same situation arose during the foot and mouth episode when vaccination was suggested; another problem is that vaccinated animals can be potential carriers of the disease. Isn't it interesting that Brian May who as I said probably has the best of intentions from a humanitarian point of view, allowed the culling of deer on his estate which he bought only a few years ago, on the flimsy excuse that the culling of deer probably roe, fallow and possibly muntjac, had been done there before he bought the estate, and he would allow it to continue until he made up his mind as to whether or not he felt the benefits to the estate warranted the cull. While he was making his mind up twenty-three mostly young animals were killed by a deer stalker. To those who know little if anything about deer culling, I would point out that a large number of young animals are taken out each year with a predominance of young bucks (you do not need very many bucks to keep a healthy population going) and a carefully assessed number of adults of both sexes, taking out the old and non-productive animals. A responsible stalker will keep an accurate record of the deer population on land which he or she shoots, not like the deer poachers who either shoot or more often tear down with lurcher-type dogs at night an unfortunate animal which happens to look towards a high powered torch shone in its eyes. Deer are particularly vulnerable to torch beams at night and will often just stand blinking at the light, until the released dogs tear it down like a pack of wild dogs on a gazelle. Brian May allowed the killing of indigenous wild deer for reasons of forestry or crop preservation. The killing of badgers would have been done for the preservation of the health of our domestic cattle herds and also the health and wellbeing of our wild indigenous badgers. How an intelligent man like him cannot see why this unfortunate course of action will have to be undertaken is beyond me; perhaps he should have taken a university course in land and animal management if he wished to increase his education. I sympathise with him in his heart-felt humanitarian feelings, but to do and allow to be done in his name what he did to those deer is contradictory to what he says about animal welfare. Recently it has been said that many thousands more deer need to be killed every year to reduce the population, ostensibly to protect the forestry and the nesting areas of birds such as the nightingale. As far as I am aware no-one has raised a voice in protest. Apparently our indigenous deer must die to protect commercial forestry interests, but not one badger must die to try to protect our national bovine herd.

* * *

Returning to TB in badgers we have a large and very old badger sett at the top of a field that we rent. When passing the sett recently I found a young badger of the year, dead outside one of the entrances. It was in extremely poor physical condition and very emaciated. I did a post mortem on it and discovered what I suspected to be a tubercular abscess in the sub-maxillary lymph gland under its jaw. Another suspect lymph gland in front of its shoulder the pre-scapular was also found. I say suspect, because without laboratory confirmation I cannot say for certain that the lesions were tubercular. I contacted DEFRA to see if they required samples for confirmation and awaited their reply. In due course I heard that DEFRA were not interested in looking at a selection of lymph glands from this particular badger; they probably have plenty of suspect badgers to look at if they are so inclined. I do not forget that my friend at Langford was last asked to look at all British mammals except badgers for evidence of bovine TB infection. As I said before if you are not looking for something you are not going to find out how common bovine TB is in our indigenous populations.

It has been said that there is no proof that the strains of bovine TB affecting cattle are the same strains as those affecting badgers. Evidence has now emerged in Ireland where investigation into badgers and cattle in a particular area showed the identical strain of bovine TB in both species.

Very recently I had a phone call from a friend who asked if I would dispatch another roe deer which had been hit by a car on a tiny little lane running parallel to and almost underneath the M5 where it goes up on stilts heading south between Portishead and Clevedon.

My friend was imminently going out but she said a friend of hers was at the scene and she had contacted a local veterinary practice to see if someone would come out and euthanase the deer. Their response was that if the animal could be transported to the practice they would then put it down, but they were not prepared to come out. There was obviously no money in it.

The lady was then at a loss to know what to do next so she rang her friend who knew where to turn under these circumstances. I arrived within 20 minutes, having negotiated the rush hour of people commuting to Bristol. There were in fact three cars in the narrow lane with three very concerned ladies. They had lifted the very large roe doe into the back of one of the cars and wrapped her in a sheet covering her eyes which under the circumstances was a kindly act to minimise stress. I lifted the doe onto a roadside bank. The lady who had supplied the sheet made her exit at this point, not wanting to witness the demise of this traumatised animal. The other two elected to stay and were able to witness the instantaneous death. I would like to think that some kind person would to the same thing for me when the day comes when life is no longer pleasant.

The second lady left and the third, the one who had called the veterinary practice to no avail was just leaving when I asked her whether she would like some venison, assuming that I could save some of the slaughtered animal. She hesitated but said that she was very fond of venison. I always feel that it is such a waste not to use what is still wholesome from such a road traffic accident. I realise of course that not everybody who happens on such a situation is in a position to adjudge what is wholesome or not, and in this particular case, I will point out why.

The local farmer now appeared to see what was going on, one of the few in the area of whom I was aware but had never met. We talked in general and the subject of badgers was raised. He enquired as to whether I shot badgers? I reminded him that they are a protected species but went on to tell him that I was only narrowly not involved with the publicised badger cull. He told me that he had only recently had some bovine TB reactors in his herd having been clear up to then. We talked of other local farmers known to both of us and some of whom I have already mentioned earlier in this book. These hard working men are at their wits end to know when their problems are going to end and are despairing that the politicians will ever get to grips with this unsavoury state of impasse.

I returned home with the deer in a bin which I keep for just such an event, to save an blood leakage into the car. There is only one thing worse than blood spilt into a car and that is milk. I found this to my cost many years ago having spilt some milk into the fabric of a car. Having skinned the deer I felt that although not emaciated in any way and allowing for an older animal which may have lost condition from calf rearing during the previous summer and autumn, she was not as well fleshed as I would have expected for her size. My mind returned to the emaciated buck which I had killed about nine months before.

Having finally removed the head-skin from this animal, I immediately saw a grossly swollen parotid lymph gland which is situated just below the ear high on the cheek. TB was the reason for this enlargement and at a stroke negated any further thoughts of any salvaged venison. The carcase was surprisingly free of damage from its encounter with the car, only a small amount of bruising around the tail was visible. Had it not been for the evidence of bovine TB most of the carcase could have been saved.

* * *

I have a friend who stalks with the man with whom I teamed to slaughter thousands of animals during the foot and mouth debacle. This friend has only fairly recently taken up the pastime. He took one of the courses for those who are wanting to take up the pastime, during the course, very superficial instruction was given on the subject of meat inspection. In fact the instructor told the group of students that roe deer do not get bovine TB! My friend had not seen an in depth post-mortem done on a deer at any point on the course, so was interested to see one done on an animal which was already known to have TB. I invited him to come and watch the procedure. I had encompassed the head and neck of the animal in a plastic bag to prevent any leakage of blood and body fluid onto my premises, even though I know that my local badgers are infected as I had previously described.

I started with the legs and shoulders, slicing finely the pre-scapular in the shoulders and the popliteal lymph nodes in the hind legs. I then removed the limbs completely leaving me with the torso and head, leaving further examination of the head until last. Opening the abdomen I found the doe was pregnant with twins, better for them to die under these circumstances than maybe later when their mother would have become more and more debilitated as the disease tightened its grip and she would not have been able to protect and nourish them. When the inevitable came to pass and she died, they would probably have to endure a slow unpleasant death by starvation; that is of course unless they were taken out quite quickly by a predator. Continuing my post mortem I finely sliced the mesenteric lymph glands and found several pinhead sized lesions. There was one lesion in the right lung and the mediastinal lymph gland was badly infected. Removing the head I found another large infected lymph gland in the throat, this being the retropharyngeal lymph node. Evidently this animal was suffering from a general infection throughout its body and probably made it feel extremely unwell and could well have contributed to it becoming a road traffic accident having maybe become slow moving.

Perhaps those who adamantly refuse to accept anything being done in the way of reducing the badger population should consider the suffering being caused to cattle, deer, cats and badgers and some sheep. The delay in taking some positive action has allowed this disease to spread and I do not think we actually know how far amongst wild deer which probably crawl away to die and ultimately are consumed by predators and carrion eaters.

* * *

The human forms of TB are returning to this country with a vengeance. The levels in the 1950's amounted to something like 50,000 cases year, they dropped to their lowest levels in the 1980's in the region of 5,000 a year. In the last 10 years the number of cases has risen, amounting to a 30 year high with a figure of 9,000 recorded cases in 2011. This represents the highest level since 1979 when there were 9,200 cases.

It has to be pointed out that during all those years since the level was at its lowest in the country, there has been an unsustainable level of immigration into our once beautiful island. Some of these have brought with them TB. It is a major cause of death in India and the sub-Saharan African countries. There is a very antibiotic resistant strain coming out of Eastern Europe and Russia thought to have originated in the Russian Gulags.

An estimate of around 70% of recorded cases in this country is to be found amongst recent immigrants. There are also many cases to be found amongst drug users and those living in a homeless situation. There are some working in the NHS who fear that this disgusting disease is spiralling out of control as it is with our cattle and badgers. The latter with too little action too late on the part of our politicians, and the former as a result of a complete failure to test all immigrants for all disease on arrival. Probably the testing of our own citizens who have lived abroad for a time when they return to this country might be a good idea. This common sense procedure would probably infringe their human rights as declared by Europe so we shall just have to continue to lose some of our indigenous population.

Young Craig White at the tender age of 21 suffered from terrible headaches and a feeling of nausea for months prior to his eventual death. His condition was undiagnosed by the medical profession because his symptoms were at first thought to be meningitis. Post-mortem revealed that his tragic death was caused by disseminated TB. Probably because the incidence of disseminated TB is rare in this country, the doctors failed to diagnose the condition but perhaps with more and more immigrants coming into the country the medical profession should be better prepared for an early diagnosis. If caught early TB can be treated with antibiotics, but if left it is fatal.

* * *

Within a week of my last recorded account of a road traffic accident involving a roe deer, I received a phone call from my great friend with whom I worked during the foot and mouth experience. He is a very experienced and dedicated deer stalker. Much of his stalking is done in Berkshire on large estates. His call was to ask me if I wanted any parts of a fallow doe he had just shot. He is not one of the new stalkers who have only seen meat inspection on the cursory courses run in several parts of the country for new stalkers wishing to be granted a firearms licence for big bore rifles. He is a fully qualified meat inspector in his own right and is very aware of the finer points of the subject. He told me that the lungs and the intestines were infected but not in this case the head as was the case in the roe deer that I had killed. He agreed to keep the offending organs for me although I shall need to pickle them before I can use them to teach my students at Langford. Because lesions had been found in two sites, the carcase of this animal needed to be rejected for human consumption.

I would like to point out how far apart these last two cases were. My friend has encountered the disease several times before in Berkshire, and as I have recounted I have found TB in several species from

North Somerset, Gloucestershire and from Cornwall. How far do the politicians want this thing to spread before they decide to sensibly address the problem?

Recently my neighbouring farmer sent two barren cows for slaughter down in Somerset. News came back to him that a suspicious lesion had been found in one of them, and by nightfall he had received a call from DEFRA invoking a movement restriction. This farmer had been closed down under TB restriction for some time previously, all he can then do is send animals for slaughter or possibly to another farm which is itself closed down using a route called 'the Orange scheme'. So called 'Orange Scheme' animals have to be kept inside and not allowed to graze on pasture. He had however been recently clear of TB and restriction had been lifted from him. He also had a bunch of store cattle ready to be sold as well as the two cows for slaughter. The decision was made to sell the two cows first rather than the stores. In retrospect, this was a bad decision because he cannot now sell the stores for at least six to eight weeks, while the suspicious lesion is being processed by DEFRA. If it is found to be a positive TB lesion, he will be back on sixty day TB testing of all bovine animals on the farm and needing two clear tests to be able to sell his store cattle. Farmers are becoming desperate for some sort of action on the part of the politicians instead of the procrastinating bulls..t which is all that has been forthcoming in the way of protecting our national herd.

Just before my deliberations on bovine TB went to be printed, I was passing through Nailsea abattoir on my way to Langford. Outside were two skips containing four best beef bodies slashed and stained blue. Enquiring from the inspectors as to the reason for the carnage, I was told that generalised bovine TB was the problem. Sadly that local farmer lost something like three to four thousand pounds worth of quality beef which was totally wasted, as was all the time and nurturing that it had taken to bring them to fruition. I thought that if he had had these animals tested before he sent them for slaughter and if they had reacted to the test, at least he would have been compensated by the Government i.e. the tax payer instead of losing everything as he did.

In summary, the kindest thing that can be done to bring bovine TB under control is to completely annihilate the colonies of badgers in the grossly contaminated areas. Any suggestion of a percentage cull is absolute nonsense and would probably cause far more long term distress amongst the survivors. The cost of slaughtering an average of 26,000 cattle a year over the last 10 years was £500,000,000 and the projected cost to the tax payer over the next 10 years if we continue as we are at the moment may be one billion pounds. This frightening forecast is only part of the story. The feeling of desperation amongst the cattle farmers in this country could see many more leaving this area of the farming industry.

I forecast in my first book that the final cost of controlling TB in this country will exceed that which it cost to bring foot and mouth disease to a successful conclusion and heaven knows that was an astronomical expense, a large amount of which could have been saved with better organisation instead of political panic.

Putting aside emotional reaction and political juggling please let us look at the science and the experience of those who have been here before, years ago, when this disgusting disease started to re-infect certain areas of the country after the initial cattle cull. I know from personal experience and that of other older meat inspectors that the incidence in cattle at the point of slaughter had diminished dramatically. This improvement continued for many years until the gassing of badger colonies in those unfortunate areas was brought to an end due to political and emotional pressure. Since then this disease has spiralled out of control and I think our chances of bringing it under control now are severely diminished and sadly far more drastic action is necessary. Whether or not the political class is prepared to grasp the nettle remains to be seen. What I do know is that the longer we delay the more draconian will be the solution. We will just have to watch this space.

CHAPTER 29

TB IN BURGERS

THE NEWS FINALLY broke over the weekend of the 29 and 30 June 2013 about the use of tubercular flesh for human consumption. The Daily Mirror, unlike most of the other papers made it front page reading on the 1 July 2013. They talk about this deception going on for at least six years; well take it from me it has been going on for far longer than that. My paper, the Daily Mail, surprisingly consigned the story to the inner pages and to my knowledge most of the other papers if they reported it at all followed suit.

I understand that TB reactors in 2012 numbered some 38,000. These cattle were destroyed at vast expense to the tax payer and the farmer alike.

My particular interest is the consumption of a huge proportion of the 38,000 carcases by the general public of this country. I have spent my life trying to protect the public health by removing diseased carcases and offals from our food chain. It has to be said that many of the diseased carcases and offals which are rejected for human food are rejected under the fairly general heading of aesthetically repugnant which in lay terms means that we would rather not eat them given the choice.

There are diseases and parasitic problems which we could eat in a raw state with no chance at all of causing us any ill effect. But when a disease such as bovine TB which has historically killed thousands of our forebears, is treated with such disrespect that our politicians and their advisors quite happily allow the general population to ingest vast quantities of infected flesh and millions of gallons of milk from cattle which are suffering from this disease prior to them being tested, this is quite another matter.

These 38,000 cattle are slaughtered as reactors, and if visible lesions are found in two sites in the body by the inspectors at the abattoir the flesh must be rejected, but if lesions only found in one site the flesh may be passed for human consumption. Are the FSA and our politicians unaware that there is something called "blood" which circulates around the body carrying with it potential problems which may in due course become visibly apparent to the inspector in the lymph glands, but prior to this nobody living can possibly know whether or not bovine TB infection has spread throughout the entire body? This reality is borne out by the fact that many cattle which have reacted to the test show no post-mortem sign of infection. Cattle which are detected as reactors will have had the last drop of milk squeezed from them prior to leaving for the abattoir.

Putting all this into perspective, there is no doubt that the application of heat to both milk and flesh will kill the offending bovine TB bacillus. Milk has been pasteurised for years, for this and other reasons. The last human death from bovine TB, which I know of, was that of a girl living in a squat and drinking unpasteurised milk from a local farm. Several of her friends were infected but were saved by the intervention of antibiotics. Meat generally is cooked, thus achieving the same sterile result, except perhaps for those like myself who prefer their steak raw in the middle.

The big question, therefore, is should we be eating tubercular flesh in the first place, with or without cooking or pasteurising? Meanwhile other diseases, such as various parasitic problems, many of which we could eat raw with no ill effect, are rejected out of hand at the point of inspection, are placed in a 'category one waste bin', are stained either black or blue and are generally treated as toxic waste. However, meat with bovine TB which, make no mistake, can affect and eventually kill a human, can be passed as fit for human food on the supposition that somebody, somewhere is going to cook it properly. You really couldn't make it up.

The next big question is of course why is this happening? The answer lies as usual with our politicians, their exorbitantly highly paid advisors and European money. European money pays for the TB testing and buys the reactor cattle from the farmers. The vast majority of these reactors are passed as fit for human consumption as I have described. Their flesh is then sold by DEFRA (Department for Environment, Food and Rural Affairs) to the abattoir owners who have won contracts for slaughtering the reactors. This can be done on a very competitive basis, allowing those selected abattoirs to sell on the flesh to various outlets probably at a very competitive rate. Most is sold to the manufacturing areas of the industry, due to the fact that most of the animals concerned would have been older milking-type animals and therefore only fit for manufacture. There are of course many young beef cattle and even calves which are destined for the quality end of the market which have reacted to the TB test. The money that DEFRA get for the reactors passed as fit for consumption is now made again available to the Government may possibly be squandered on other just as poorly thought-out schemes. If, as most members of the public thought, all these reactor cattle were slaughtered and burnt, the cost would escalate astronomically, as in the days of the 'foot and mouth' debacle, which I predicted in my last book would ultimately be overtaken by the cost of the bovine TB scourge.

This has been swept so conveniently under the carpet by several generations of politicians in the name of political expediency and the ever present worry about votes. Nothing has been done for years about the badger problem in case the town dwellers who tend to side with the 'don't kill any badger brigade', might take their revenge on any political party who had the temerity to actually do something positive about the problem.

To return to the media coverage there was a huge outcry when horseflesh was found to be widely sneaked into the manufactured meat products on sale here in the UK and abroad, shouldn't we be more concerned about Bovine TB?

With the horsemeat scandal, probably the horses concerned were healthy, apart from a few drugs hidden in their flesh. The media exposure went on for weeks. The subject of TB flesh being consumed over many years lasted exactly one day. The reasons for this could be that the public has become complacent about being conned at almost every turn and aspect of their lives, or perhaps pressure has been brought to bear in certain quarters to kill the story in the name of national financial interest. Or just possibly we must not upset our European partners who in several cases took a pasting over the horseflesh scandal, and it has been decided that further scandals must be kept under wraps. I found it quite amusing to hear the hierarchy of the Food Standards Agency bleating about checks being made at the point of slaughter. I wonder if the people polishing their trousers and skirts are really aware of the dumbing down of the meat inspection service, some dedicated exceptions apart.

The FSA are even talking about a just visual inspection of pigs that have come from farms where they have been veterinary inspected alive and obviously must be OK to eat. We are not dealing with nuts and bolts or potato crisps here; we are dealing with individual animals with individual problems which are sometimes quite difficult to find.

The way we are going reminds me of a time when, while working at Gordon Road, a high ranking woman from the Bristol Treasurers' Department came to visit the abattoir with a view to cutting costs to Bristol City Council. Having watched the various procedures in place for inspection she came out with a classic comment which I will never forget; "Couldn't you only inspect every other one?" To which I responded "I cannot believe that I heard you say that"!

CHAPTER 30

POISONS

THE INDISCRIMINATE USE of chemicals worldwide is very, very worrying. Big business, of course, does not want to hear any voices of dissent, because that is bad for business. You hear of safe levels of contamination. Then you hear later that those levels were not safe at all. You only have to look at the disaster of D.D.T which can still be found in the environment from decades ago.

It seems that all crops have to be continually sprayed with both herbicides and pesticides. It makes me wonder how we ever survived before these modern poisons were invented. Apparently you cannot even dig potatoes anymore without spraying off the haulm with a herbicide. Corn has to be sprayed with herbicides pre-emergence and two or three times after emergence to kill off any green stalks, when the harvest is due. It is again sprayed with pesticides for aphids and other pests. Not only are we eating chemicals in our vegetables but also in the flesh of food animals. In fact various flesh and organs are routinely tested for residues.

All of a sudden the birds I remember as a young man are no longer there. Skylarks were constantly singing in the valley where I live; I have not heard one now for 20 years. Grey partridge were common in the valley 40 years ago. They disappeared, and even though, when keepering a small shoot in the valley we re-introduced grey partridge, and very soon in the annual bird spotters' reports it was noted 'grey partridge return', I wonder if we were ever given the credit for that? Sadly the re-introduction soon petered out with intensive agriculture killing all the food insects for young partridges, together with the monoculture of various corn crops, an alien environment for the very young birds.

You see advertisements on the television to spray all your pests in the garden and all your weeds. Are we now too tired to dig up the weeds and squeeze the caterpillars on the cabbages? I knew an old market-gardener friend long dead now. As a boy his father would pay him 2d for 'a hatpin' full of skewered slugs, gathered on a damp night, with the aid of a torch. Can you see the young of today earning their pocket money in this way? They are probably too busy on the mobile phone or surfing the internet.

No wonder that I very rarely see a song thrush anymore, when they used to be very common garden birds. Our bees are dying, particularly in the countryside where everything is blanket sprayed. Interestingly bees are increasingly being kept in towns by people wishing to get back to nature as far as they can, even keeping bees and chickens on roof gardens. These seem to be flourishing, which tells me that some people at last are becoming aware of the danger of chemicals in their gardens. I only hope it is not too late!

All these horrendous poisons in our environment we are told by the politicians are perfectly safe. How about if they mix together? How about if the farmers over mix them? How about we stop using them?

Perhaps if the politicians had taken notice in 1951 when Professor Solly Zuckerman chaired a working party which produced a report for the Agricultural Minister 'Toxic Chemicals in the Environment'. Whilst discussing organo-phosphates Professor Zuckerman repeatedly warned about the danger of chronic effects. Even though the Professor was the Government's Chief Scientific Advisor his warning words were completely ignored (too much big business was involved). There have been so many dissenting voices since, by the more discerning and responsible scientists. All falls on deaf political ears.

My particular interest in all of this is B.S.E. Everything to my mind points towards the cause being the pouring of organo-phosphates over the backs of cattle to rid us of the Warble fly. Allowing for a predisposition of certain individuals to react to those particular organo-phosphates, I have seen to my own satisfaction how this is definitely the case with humans. This leaves me in no doubt that the same thing applies in cattle, which is why only some of the herd developed B.S.E. It is also interesting to note that B.S.E has virtually disappeared from the national herd, which coincides with the withdrawal of organo-phosphate 'pour ons' for Warble fly destruction. We shall of course never be told the truth about all this. But will the politicians learn that when they get it wrong (and everybody gets things wrong) they should hold up their hands and say sorry. We the public at large, would have far more respect for them, instead of like me, despising them for running for cover like rats leaving a sinking ship.

It seems we have arrived at a level where we use our soil which is the basis of everything, as a medium for growing things in rather than allowing the natural regeneration of fertility, as has happened for thousands of years. We now use artificial fertilisers mostly coming from by-products of the oil industry, to force each crop into unnatural abundance. What happens when the oil runs out? In the meantime there is excessive use of nitrates which run off the land and poison our natural water resources. We used to fallow the fields and fertilise with natural manures, it might not have been so dramatically productive, but it was definitely sustainable. Does not this in itself tell us we are taking from nature at an unsustainable rate? I keep hearing talk about 'growth' in all aspects of our lives, how much 'growth' do they think that the planet can take before it takes revenge on us, the human race, for what we have done to it?

With so many countries now having nuclear capability, it will not be long before some fanatical regime presses the unstoppable button worldwide. This is a sad indictment of the human species that we would wish to annihilate millions of our own species. Nuclear power in itself cannot be justified, we should not make something that we cannot destroy, and have to leave as a toxic legacy to future generations. Thousands of unsightly windmills which have apparently failed miserably in America and more recently in this country are not the answer. Surely it would not be beyond the clever scientists to invent something to harness the power of the tides without environmental impact. Whatever we do they will carry on going in and out twice a day.

Are we so disenchanted with this beautiful planet of ours that we have to abuse our bodies with the addictive substances of drugs, alcohol and nicotine? I do find it offensive that those of us who are able to resist temptation to pathetic drug addiction, are expected to fund the Government's rehabilitation projects which seem to have very little success. If people are that stupid with all the publicity that is available today then to my mind they deserve all that they get.

Something which has concerned me increasingly over more recent years, has been the use (in feed and by injection) of various antibiotics and growth promoting additives in the rearing of food animals and birds. This is necessary because we have such huge expanding populations not only in this country, which is bad enough, but also globally. That in itself is bringing us closer and closer to disaster, and is certainly not a sustainable progression for the human race.

To satisfy the huge increasing demand for more food, not only the British but also the world's farmers have had to intensify their operations to try and keep up. Intensification brings with it a whole new set of problems. It seems that as soon as we start to mess around with Mother Nature she fights back, sometimes on a grand scale.

My particular concern is the food animal and bird industry. We cleverly thought that injecting growth promoting hormones into cattle would achieve a saleable carcase in much less time that it normally takes, therefore less food would be necessary to feed the animal concerned and therefore more profit could be made. What is cancer but cells which have gone berserk? What are we doing eating flesh which has been polluted in this way? It was not very long before these growth promoters were banned in this country, obviously somebody somewhere realised that perhaps it was not a good idea to inject these substances into our beef which in due course would be consumed by the human population. However I did hear that it could still be obtained either on the internet or from other countries.

Pigs soon became intensively produced causing them all sorts of health problems. Whenever large numbers of animals are kept living on top of one another as they are, inevitably there is a build up of bacteria in the hot foetid environment in which these animals have to live. Successive generations of animals were open to various infections left behind by their preceding counterparts and various syndromes of intensive farming induced disease began to appear. I have lived and worked long enough to have witnessed the change from traditionally kept pigs through to intensive production which it has to be said in many cases has improved over the years. But some people have opted out into traditionally kept pigs by way of hobby-farming and in some cases large scale traditional farming mostly for the organic market. Traditional diseases which I remembered from my early inspection days, were appearing again with the return to more natural farming methods. Predominantly these diseases took the form of parasitic problems which had for years become rare as a result of routine worming during mass production, something the hobby-farmers probably for the most part were blissfully unaware of. I welcomed the return of these indigenous parasites because it allowed me again to show my various students not only the parasites themselves, but also the various visible body damage caused by them. However if the infection was so bad that it was causing the animals themselves or the owners' pockets preventable damage I would take it upon myself to ring the owners concerned and advise on the necessary corrective treatment. If however the infection was of a sustainable level and causing the animals concerned no particular problems, I would let nature take its course and have a regular supply of teaching specimens.

The poultry industry has a lot to answer for keeping these birds in conditions where they can hardly move and exercise their little bodies. This is known to many, but was brought to the wider public's notice by the famous chef and television personality Hugh Fearnley-Whittingstall who was so affected by the horrendous conditions under which these little birds were reared and kept, that he was seen to shed a tear while talking on one of his programmes. By speaking out as he did he was castigated by the big supermarkets, which were as ever terribly afraid that their businesses would suffer as a result of the publicity. Hugh also graphically showed the 'by-catch' in the fishing industry being dumped at sea out of the public gaze, another madness we can thank the European Commission prats for. All that food and life wasted for a scheme dreamed up by the 'know nothing' shiny trouser brigade.

I had the good fortune to meet Hugh at the Nailsea abattoir, when he came there to have a pig killed in association with one of his excellent self-sufficiency programmes. He is just the same man as we see on the television, not only showing his skills in the kitchen, but also pointing out how wasteful a society we have become. He like me, believes that whenever we take an animal's life we should use all that can possibly be used of the resultant carcase and offal. He also likes to source food from the wild, something that we who are historically "hunter gatherers" have been doing since the Stone Age.

The potential poisons in the intensive poultry industry start before the birds are even born, when a few days before the eggs are due to hatch they are injected through the shell with various antibiotics and vaccinations to prepare the unborn chicks for the intensive environment into which they will be born. Without these preventative medicines they would stand little chance of survival in a world of salmonella and campylobacter to name but two. Apparently around 40 billion chickens are eaten by humans every year around the world and the vast majority of these are intensively reared. Due to the horrendous overcrowding in their rearing houses, the incidence and danger of disease spread does not bear thinking about. Just taking the campylobacter organism, in 2012 it infected 580,000 people in this country alone, all thought to be in connection with poultry. 18,000 were hospitalised and 140 died. Using antibiotics as a preventative measure in our table poultry, allows nasty little bugs like campylobacter to mutate to accommodate the currently used antibiotics. The danger of this is twofold: the resistant bugs will probably kill millions of chickens, but worse will be the cost to the human race that could die in vast numbers with little in the way of new antibiotics in the pipeline to kill the resistant mutated bacteria. The first could be a financial disaster for the producers. The second could keep the population down, but do we want to die in large numbers through greed and ignorance?

Could the wide use of antibiotics in the rearing of food animals be responsible for the so called super-bugs which seem to infect our hospitals today? These superbugs are killing increasing numbers of sick and older people whose resistance is decreased by age and infirmity, but they can also strike down otherwise healthy people; all in the name of cheap food available today.

It would appear that eighty percent of antibiotics used around the world today are used in connection with our food producing animals, roughly fifteen percent on human patients in the community and only about five percent in the hospital environment. There are strains of E Coli which are becoming increasingly resistant to a type of antibiotic called Cephalosporins which are very important as almost a last chance in saving life when all else has failed. If things go on as they are many of us are going to die painfully as a direct result of treating the life saving gift that Fleming gave us with contempt as we have been doing for decades. Instead of using Fleming's and many others' discoveries as a last resort lifeline, we have been dishing them out like sweets for the slightest little ailment. In fact the public at large have become so complacent on the subject that they feel thwarted if they do not come away from a doctor's appointment with a course of antibiotics for some comparatively minor ailment. Very often as soon as they feel better they fail to complete the prescribed course of tablets. The bacteria concerned which may only have been weakened but not necessarily killed have a better chance of mutating to resist that antibiotic, so that next time, a stronger one has to be used. We are getting to the point where we are fast running out of stronger ones. Surely it is better to allow our bodies to build up their own resistance to these nasty little bugs, and only use antibiotics as a last resort. This should apply not only to us as patients but also to our food animals which it seems are bombarded with them from birth to keep them alive in the horrible intensive conditions we have imposed on them, all in the name of supplying cheap food for what is becoming an unsustainably large human population.

As if poisons and additives in the flesh of our food animals and birds is not enough to cause all sorts of possible problems for our bodies to assimilate with who knows what long term implications, it is now revealed that getting on for half of our fresh fruit and vegetables are contaminated by increasing quantities of pesticides. Top of the list seem to be oranges with pears, grapes, pineapples, apples, raspberries and carrots closely coming along behind, but the most frightening thing of all is that flour and therefore our staple diet of bread is also heavily contaminated. Many of these items come from abroad where it has to be said in many places human life is very cheap. I once saw a programme where cut flowers destined for the British market, the sort of thing you see on every garage forecourt, were being grown in poly-tunnels in the desert and while the impoverished workers were tending the plants the owners of the project were spraying the crop with pesticide. If they are prepared to do that to their own people for monetary gain, how much more likely are they to heavily spray fruit and vegetables destined for export and to be consumed by people in foreign countries? I, and I expect many others, think that a good wash before consumption will remove the poisons but even peeling does not necessarily remove them, particularly on things as staple as potatoes which in many cases are eaten in their skins. Something called carbendazim seems to be one of the most commonly found substances, which has been linked to birth defects both in animals and humans as well as evidence that it is also a carcinogen. Other pesticides have been linked to the demise of bees, without which we have lost the best pollinating insects known to man. Close to home my daughter has kept bees for a number of years. One by one her hives diminished and eventually died out, not I may add due to the disgusting little varroa mite which like many other foreign imports has done irreparable damage to our once cherished and beautiful little island. I am thinking now of Dutch elm disease, Ash dieback disease, grey squirrels, brown rats, Japanese knotweed, Zander fish, American mink and South American coypu, the latter being the only one to my knowledge that we have been able to eradicate. There are many, many, more which have been introduced over generations to the detriment of this country.

It took the Irish scientists to bring the news that washing did not clean these pesticide poisons from the surface of our fruit and vegetables. Our own Food Standards Agency, as with the horseflesh exposure, were still playing catch-up to the Irish and only managed to warn us about the problem in 2012 when the problem had been increasing for years.

Fortunately I have a large garden and I am able to grow much of our vegetable needs. Many are not so fortunate and have to rely on the powers that be to see that our food imports as well as all that is produced at home is not only wholesome but also it is not going to slowly poison us. They have failed us miserably.

Recently in Marlborough somebody, probably in all innocence, must have poured maybe as little as two teaspoons full of pesticide down a drain. This particular poison Chlorpyrifos by name is used by many gardeners to eradicate predating insects on garden crops. The result was that a ten-mile stretch of the river Kennet was contaminated and all the small shrimps and various fly larvae were wiped out over all that distance. These small invertebrates are the basis of the river's eco-system and therefore the fish, many of them trout which we eat, as well as the predators like herons and otters were inevitable affected with who knows what repercussions. At the height of the contamination local people were warned not to get the river water on their skin and definitely not to eat the trout caught between the towns of Marlborough and Hungerford.

An environment minister who lives by the Kennet river further down towards Reading, has apparently instructed his people in DEFRA to investigate as to whether Chlorpyrifos can be banned for domestic use. Surprise, surprise, isn't it amazing that when our Honourable Gentlemen are personally involved with something like this how suddenly something has to be done about it? Isn't this too little too late? How can something as toxic as this can be used at all whether domestically or not? This particularly toxic substance has already been banned in some parts of the world, so what the hell are we doing allowing it to be used to poison our once lovely country?

CHAPTER 31

MADNESS IN COURT

COMPARATIVELY RECENTLY WHAT I consider to be the ultimate in bureaucratic madness has occurred at a small abattoir in Somerset. This manifested itself in the shape of a particularly arrogant and obnoxious Spanish vet, who although initially working at the plant as an Official Veterinarian, was in fact one step higher on the ladder, known as a "lead vet".

The first confrontation I had with him concerned the stunning and bleeding of sheep. I only work occasionally as a casual employee of then the Meat Hygiene Service, now the Food Standards Agency, (these agencies seem to come and go like the wind blows). On this particular day I noticed that the slaughterman cutting the sheeps' throats was not carrying out a process known as "pithing". This involves in the case of sheep the severing of the spinal cord which is achieved in three possible ways. The first, and probably the most practical, is having cut the sheep's throat severing both carotid arteries and both jugular veins, to insert the point of the knife between the atlas vertebra and the skull. The second is to cut the head nearly off and just leave it hanging by a piece of skin on one side. The third method is to bend the head back after throat-cutting and to break the neck. This method is easier to achieve if the animal is stunned on a cradle rather than hanging upside down on a leg chain.

The reason that this process had been performed for generations was to ease the passing of the sheep in its final moments, working on the same principle as in the good old days of human execution, when the guillotine was considered the most humane way to put some-one to death, instantaneously severing the blood vessels and spinal cord. In this country our own method of capital punishment by hanging, (now sadly missed) when properly performed by Mr Pierrepoint and others achieved the same instantaneous result.

When I questioned the slaughterman concerned, he told me that the new Spanish vet had told him he must no longer carry out this procedure. I was concerned about the welfare implications of this new directive and went to speak to the vet quietly in his office. In a dismissive manner he told me that pithing was illegal. I asked him if in fact he was a Spanish gentleman and immediately on the defensive, he replied in the affirmative, and the following conversation ensued:

"Is Spain part of the EU?" – "Yes"

"Is Great Britain part of the EU?" – "Yes"

"Is this illegal pithing European law?" – "Yes"

"Is bull fighting still allowed in Spain?" – "Yes, but I don't agree with it"

"Do they mutilate the animal by driving barbed coloured sticks into the shoulders?" – "Yes"

"Do they then gouge its shoulder muscles with lances on long poles from a padded horse?" – "Yes, but I don't agree with it"

"Having weakened the bull to make it drop its head and make it safer for the matador to approach it, do they then try to dispatch it with a sword?" – "Yes, but I don't agree with it"

"If after all this, the tormented animal isn't quite dead, do they then with a short dagger sever its spinal cord from behind the horns between the atlas vertebra and the skull?" – "Yes but I don't agree with it"

"Well then, Mister, can you tell me the difference between that bull in Spain and a lamb in a Somerset slaughterhouse if we are all in this EU together?"

He did not reply, whereupon I returned to my duties in the slaughterhouse. He very soon followed me, and very unprofessionally in front of the plant staff accused me of insubordination, much to the amusement of those listening.

He was later to accuse me in my absence at a high powered meeting of the Meat Hygiene Service of having confronted him unprofessionally in front of the plant staff whereas the complete reverse was true. This showed him to be prepared to lie to further his own ends. Unbeknown to him a very good friend of mine was present at this meeting, and having already heard my version of events, challenged the Spaniard as to who confronted whom in front of the abattoir staff. Becoming aware that his misrepresentation of the incident had been discovered he refused to answer the question. Questions were asked by those present as to who this insubordinate inspector might be. My friend told them that this man is probably the longest serving meat inspector in the country and has probably taught more meat inspector, environmental health and veterinary students than anyone else in his position, and in fact trained him as well some 12 years previously. Turning to the Spanish vet he said "I would suggest that if this man has something to say on a particular subject you should listen, and you might possibly learn something!"

After this incident I was interested to know why it was that a process that has been used for generations had suddenly become illegal. I enquired from a friend of mine at Langford Veterinary college, and he told me there now had to be time lapse after stunning and cutting an animal's throat, before any further process can be undertaken. This time lapse is laid down in legislation as being 30 seconds in the case of cattle and 20 seconds in the case of pigs and sheep. He pointed out by severing the spinal cord the animal became paralysed and it was not possible for an inspector or vet to adjudge the efficiency of the stun. My rejoinder to this explanation was that if there was concern about the efficiency of the stun, the inspector or vet should watch the operative and see that he was doing his job properly because by the time the throat had been cut from ear to ear, if the animal had not been properly stunned, it was too late anyway. You could not sew it up and say "Sorry, we will try again!" and of course if the slaughterman happened to be of the Muslim or Jewish faith the animal would not have been stunned in the first place.

This same Spanish individual not only tried to belittle me and my experience but also adopted a very intimidating and bullying manner towards the slaughtering staff and the two resident meat inspectors. This included standing oppressively close and criticizing continuously a process which has been proved over the years both visibly and bacterialogically to produce a very acceptable carcase of meat. It is not in any case part of his job to direct the plant staff and if he has a point to raise he should take it up with the plant owner.

One of the slaughtering staff, a highly skilled man whom I have known for 45 years was so incensed by this oppressive behaviour that he wrote to his MP, Dr Liam Fox complaining about the situation, and asked me to write in support of him and the following is an extract from this letter.

"The regular inspectors of the plant I have known for many years and they are both very experienced and skilful at their job. One of them, however, on one occasion I witnessed was so stressed by this man's bullying that he was sent home to recover his composure. On this particular day I became aware of this man's complete inability to perform one of the primary duties of an official veterinarian which is the visual examination of animals held in the lairage prior to slaughter for possible disease or infirmity problems. He failed to see a bovine animal which was very evidently in a fevered condition. On post mortem examination by me this animal was found to be suffering from a very acute and diffuse peritonitis and was totally rejected for human consumption, but only after it had been slaughtered in the middle of the bovine kill. When I questioned the vet as to what he had seen in the way of symptoms in the lairage he said he had seen nothing. However, subsequently the younger son of the abattoir owner, who had been in the business for less than 12 months, told me he had seen an animal looking very hot. When I questioned him as to whether its coat had a dew-like appearance he said that it had. This animal should have been slaughtered at the end of the kill in an effort to prevent the spread of pathogenic organisms onto adjacent carcases. This incident told me that this man was not fit for purpose but latches on to trivia and in doing so tries to bully those who know more about the meat trade than he will ever know."

Before the veterinarian intrusion into the British meat trade ante-mortem inspection was carried out by the meat inspector, and I have ante-mortemed millions of animals. I am now no longer considered competent to carry out this task. Interestingly, this has only been the case since we joined the EU. These days mostly young, very inexperienced, mostly foreign vets perform this part of the procedure. As I have been entrusted with trying to teach these young people in a very short time this would appear to be an anomaly.

It has to be said if these young vets do manage to see a problem in the live animals I have yet to see them communicate the condition with a meat inspector who is doing the post mortem examination. Surely it is better for the man who is doing the post mortem inspection to do the ante- mortem inspection as well for continuity.

The third confrontation I had with the same Spanish vet, concerned the washing of sheep carcases. For many years it had been considered perfectly acceptable to wash carcases for the removal of visible contamination, be that faeces, earth, hair or blood. No mention is ever made of the invisible contamination which occurs as soon as an animal's skin is removed. This particular vet decided in his wisdom to forbid the washing of sheep and that all visible contamination, however miniscule, should be removed by incision. Not one of the many other vets attending the plant over the years since Europe dictated we needed a vet in the slaughterhouse ever questioned the washing process. It has to be said, however, that it is very undesirable for a carcase to be made wet more than is absolutely necessary, as warmth and moisture encourage the rapid multiplication of bacteria which are ever present. Nevertheless visible contamination has to be removed for aesthetic reasons and the best way to do this is a combination of incision where feasible and the application of a high pressure high temperature water jet to remove the infinitesimally small flecks which do inevitably occur however careful the operative was when dealing with animals that spend their lives in a world of rain and mud and faeces. At the end of the day, if the inspector is good at his job and the animal is healthy all contamination is on the external surfaces, and whether it be on a joint or a chop or a steak the external surface is exposed to very high temperature while cooking, which will kill 99.99% of bacteria, very few of which would be harmful anyway. Contrary to popular belief that all meat is a high-risk food, the high-risk areas are cooked meats which may have become contaminated and are not going to be cooked again, reheated meat which may have been contaminated and then not heated sufficiently, and reconstituted meat such as burgers which turn the outside which may have become contaminated to the inside and then not cooked through. E coli is now commonly known particularly in America as "the burger bug".

I would like my readers to consider for a moment the methods employed by Ray Mears, a survival expert who appears on television, to cook wild- caught meat, such as rabbits and deer in an outdoor setting with none of the usual utensils. He cooks either on a spit over an open fire or else wrapped in leaves and buried in hot ashes in the ground. In both cases the meat is exposed to all the local contaminants but both Ray himself and those he invites to partake of these delicacies all look extremely well. I'm a great believer in the old adage that you have to eat a peck of dirt before you die and years ago this probably happened, which was a good thing because it gave people a natural immunity to the bugs in their immediate environment.

Today however we strive to live in an aseptic environment which leaves us and particularly our children very susceptible when encountering by accident some of the very common food poisoning bacteria. This is especially evident when holidaying abroad, and people almost seem to accept as inevitable that they will suffer from some combination of sickness and diarrhoea from either the local food or water.

It is generally accepted that farm children who are exposed from an early age to farm animal's faeces and urine are more healthy than their city counterparts. There have recently been cases where urban children have visited so -called "petting farms" where they are encouraged to handle various animals, and as a consequence have become ill and in some cases extremely ill.

Having said all this, every effort should be made to produce a totally visibly clean carcase from the abattoir.

Returning to the question of washing carcases, the vet decided the Food Standards Agency should take a prosecution against the abattoir for washing sheep, even though at his behest washing was suspended until the matter was sorted out. Eventually the matter came to the Magistrates court, and after two days of legal argument the case proper began. I was asked to appear as a witness for the Food Standards Agency. When cross examined by the defence solicitor I could only explain the situation as I saw it. The law states, that any visible contamination must be removed by incision or any other suitable means. This quite clearly could include jetting it off with a very hot high pressure hose. In the case of cattle heads which are sometimes contaminated by quantities of stomach content regurgitated at the point of slaughter and which can block the nasal cavity and the areas at the back of the tongue and the epiglottis, the law states that this should be washed away to cleanse and facilitate inspection; indeed special cabinets are produced for this very purpose.

I pointed out in court that it is difficult to understand how anyone could possibly interpret the relevant legislation as forbidding the washing of carcases. However, given the lack of training and practical experience sadly common amongst the officials in the meat trade these days, perhaps it is not surprising that they latch onto trivia to justify their existence and are incapable of seeing what is most important. As the saying goes:- "Rules are made for the guidance of wise men, and the obedience of fools." Common sense must prevail and in this particular case when encountering a lump of semi liquid faeces you sensibly remove it by incision rather than hit it with a pressure hose and spread it everywhere.

Fortunately the magistrates were able to see through this charade of bureaucratic madness and decided there was no case to answer. Perhaps those concerned in this particular fiasco may possibly realise that it is not always a good idea to rush into prosecutions. It has been my experience over many years that people respond well to a little bit of praise.

This became particularly evident when I spent some time inspecting butchers' shops in the north Somerset area. When first entering a premises, clutching a white coat and hat and a notepad, I would introduce myself as a health official from the local authority and as such I would viewed with some suspicion by the owner or manager. I would start my inspection and try to find something to praise, be it a procedure or the general display of goods, having already spotted something which needed attention. I would then speak to the owner/manager and ask if the good points were his idea. His response was usually very guarded and defensive, but if in the affirmative I would then tell him that it was good and I liked the idea. In most cases he would visibly thaw and it would be then that I would ask for some usually quite trivial works to be done – maybe a couple of cracked tiles behind the block which needed replacing. I would then say that I would be in the area in a couple of week's time, and if he could get the work done by then it would save a bureaucratic paper chase. Nine times out of ten on my return the matter would have been amicably sorted.

In today's world the rush to use the big stick as a first resort rather than a last resort is too often seen. It is very important to build a relationship of trust and mutual respect between officials and the trade itself, but this is not always possible due to over-zealous inexperienced officials and those in the trade who are determined to flout the law.

CHAPTER 32

PROBLEMS FACING THE WORLD – POPULATION

FROM MY POINT of view the world as we know it is close to breaking point. The biggest problem as I see it is over population by the human species. We see in nature, when any species over populates an area, that nature retaliates by thinning out the over population by disease or starvation.

Look at the Black Death which wiped out 50% of the population of just this country. The influenza outbreak at the end of the First World War killed more people than died in the war itself. There have been several more influenza outbreaks since which killed many thousands of people worldwide.

Bird flu looked as though it might potentially take out millions of people. Similarly swine flu looked ominous but neither really took off. HIV accounted for many people in Africa in particular, but there was some speculation as to whether it was a natural phenomenon or as a result of a man made experiment in Africa which went wrong.

With our superior intelligence one would think that we could see the folly of overpopulation. In the under developed countries with little if any family planning, huge populations develop which are sustainable as long as the food supply is sufficient. However as we have all seen over the years crops fail with droughts, or floods or pestilence. Vast numbers of simple people die miserably particularly the young and the old. Horrendous pictures on the television pull at the heart strings and lead more affluent populations with surplus food supplies to bail out the beleaguered. Hard as what I am going to say would appear, giving this sort of support at these hard times only allows many more to survive than nature intended. One has to look past the agony of the individual towards the greater good. If this generation survives to breed the problem is only exacerbated. When the next drought hits far more people will die than would have before.

I am very aware that my sort of thinking will not be accepted by many, but I would ask my reader to bear with me and then be honest in the conclusions that they might come to. These under developed countries do not take from the world's natural resources at remotely the same rate as do the developed nations. We in the west and America and soon China take at an alarming rate from the world's resources. Oil supplies, worldwide fish resources, timber from natural forests, bits of endangered animals (rhino horn, tiger bones etc) to name but a few.

One baby born in the west will probably take more from the world's dwindling resources in its lifetime by many many times than one born elsewhere. It has to be said that huge Third World populations albeit taking little from the planet in a material way but by virtue of needing land to live on push natural wildlife on that land towards extinction.

Just take the isolated island of Madagascar, where the desperately poor population is increasing at an alarming rate. Ninety percent of the natural forest has been destroyed. The island is home to many many types of lemurs which are unique to the island, and loss of habitat to the human population and their desperate need to grow more food has driven many of these delightful animals to the brink of extinction. Families on the island have an average of ten children each and that sort of reproduction cannot be sustained.

Taking just this country, our population has soared in the last 50 years. A large portion of this is due to unchecked immigration and we are now told by the politicians that we should be pleased to live in a multicultural society. I don't remember being asked if I wanted to. Looking around our major town centres today you could be in any major city in the world India, Pakistan, Africa, Jamaica or any eastern European city. Is that what indigenous Brits want to see happen? We used to be proud of our culture and heritage but it seems now that anyone from around the world can claim British citizenship within five minutes of entering this country. But this is all part of the Eurocrats plan to make us lose all our sense of national pride and identity before we become part of one of the planned Euro zones. Come on people take the blinkers off your eyes before it is too late! Most of the mass immigration was during the Labour years, and the Coalition promised to reduce the rate to less than 100,000 a year. Another broken political promise; why do they bother to promise anything especially when Europe can veto any decision we might want to make for ourselves.

Of course these official (probably massaged) figures are only part of the story. What about the hundreds of thousands living here illegally? I remember some years ago when my daughter was returning to this country on a cross-channel ferry there were some young men leaning over the rail. They were throwing their passports and relevant documents into the sea. My daughter asked why they were doing this? The reply was that they could not be deported back to their own country because nobody would know where they came from. So they could not be repatriated. Our national borders have become a joke!

Some 20 years ago there was an accident on the M5 near Taunton, it involved a large articulated lorry carrying boxed food products. When the police looked into the damaged trailer, there were some frightened illegal immigrants hiding there. After some discussion, they were told to run away across the fields. This at least saved the time wasting paperwork for the police, who knew that they would probably be allowed to stay in this country anyway.

The once cherished birthright counts for nothing today. The birthrate amongst immigrants is far higher than amongst indigenous citizens. Of course the more children you produce the more money you can claim in child allowance amongst other allowances provided by the long suffering taxpayer. Of course immigrants are not on their own in draining our national resources but it would seem only fair that newcomers to the country should have to prove good work ethic over many years before even being considered for benefits of any sort. It seems that what many of us have been paying into for years is now the automatic right of every newcomer that comes along.

Having said all that we have an increasingly prevalent problem with our own work shy feckless underclass of totally dependent people sometimes over several generations, who breed like rats, so that they can claim without any sense of shame what the rest of us have to work hard for over many years. It begs the question why? All these lazy parasites should be made to work before any immigrant is considered for work in this country. When all these layabouts are working, then, if there are any jobs to spare, we could allow in a few outsiders for a limited period only, which could be reviewed every now and again as the circumstances of the country change.

* * *

Human greed can manifest itself in different ways. Despicable despots who rise to power on the backs of their fellow men are usually totally heartless.

Look at Hitler, Mugabe, Mubarak, Hussein, Gaddafi, Blair and Assad, to name but a few. There are many many more whose crimes against humanity very often go unreported. It is very rare to find someone who rises to a position of high office who has not dipped their bloody fingers into human misery to obtain and retain their positions of power.

Behind these despots lie the very often faceless supporters, who usually for financial gain keep the despots in power for their own selfish reasons. Much of big business falls into this category; you only have to look at Murdoch, and the can of worms that his exposure has opened up.

Then of course there are the blatant thieves who never get prosecuted, we are now talking about the irresponsible greedy bankers who we entrusted with our hard earned cash, which they unashamedly squandered to further their own ends until their cavalier disregard for other people's property came crashing down around their ears. Unabashed, they more or less carry on regardless giving totally undeserved bonuses having 'drawn a line' under what they have done in the past and 'moved on'. They should be made to repair their irresponsible damage before the word bonus is ever mentioned again.

Our culture today is one of greed and selfishness and surrounding oneself with mostly frivolous modern technology which although amazing in its invention causes terrible damage to society at large and the young in particular. Little thought was given to the repercussions when letting the World Wide Web loose. When you hear stories as I have read in the paper only today of a boy of 11, raping a girl of nine because he had been viewing pornography on his computer, and thought that to have sex with a little girl would make him feel grown up, you realise we have stolen our children's childhood!

It would seem to me that the only humans to have got it right are those who have luckily for them not been tainted by the modern world. I am talking about the primitive tribes in South America, the Bushmen of the Kalahari, and a few others around the world who have remained in isolation. They take from the planet just what they need to survive and no more. They reproduce enough to keep their populations steady. They have no official religions to dictate their lives, although many attempts have been made to bring them into line by particularly the Christian faiths. Why do these religious people find it necessary to think that they are right, and anything else is automatically wrong? They even send people round to your house to convert you into becoming a Mormon, a Seventh Day Adventist, or a Jehovah's Witness. It only shows an underlying insecurity, if they think there are pockets of dissent out there.

My grandfather was a very kind simple man, a carpenter by trade, and one of life's gentle-men. He was inveigled into becoming a Seventh Day Adventist. It was not until after his death, that we found how much he had been co-erced into giving to his church 'for good works abroad'. I found money secreted away for periodic collection by a representative of the faith, but the last envelope had not been collected due to his demise. He led a very simple life in a cottage he built for himself and his wife in Pembrokeshire. They had no modern amenities, water from a natural spring and external earth toilets were the order of the day. A little bit of money spent on themselves would have made their lives so much easier. The amount of money they had scrimped and saved to pay into the religious coffers amazed me. These bloody parasites should be ashamed at what they did to that lovely old couple.

CONCLUSION

I CAN LOOK back over the years and think to myself how fortunate I have been to have enjoyed my working life; many have not been so fortunate. However, I look at today's world and would *definitely* not choose to do the same job again. The reasons for this sad pronouncement are also the cause of what my wife and others call my "rants" about various aspects of the world today. I would go further and say that my generation has lived through the best times.

In today's world the feeling of camaraderie with one's fellow man is largely absent. That is of course a sweeping statement. There are obviously many, many caring and decent people out there, but from the political class, as they like to be known, down through big business and banking, there is a culture of deceit and greed. There is also a vicious element in society today which I do not remember as a young man. You only have to look at our inner cities to witness a dog-eat-dog existence.

Youngsters of today are exposed to the bad side of life from a very early age, from the television and the internet. The latter invention, although an incredible human achievement and a huge advancement of knowledge, has stolen our children's childhood, and many are so dependent on their phones and computers that they can hardly converse with adults and are completely unaware of the practical, and to my mind, more important side of life.

There are still those who will not even consider entering this cyberworld and I personally will not have a computer in the house. Sadly, younger people have to comply, to keep up in today's world.

It saddens me that the legacy my generation has left to children and grandchildren coming along behind is not good. As my readers may have noticed I entirely blame the political class over many years for what they have done to this once lovely and highly thought of country. For their own ends, these politicians have destroyed our nation by taking us into the "United States of Europe", and by destroying our Britishness by allowing all and sundry from around the world to come and exert their influence, without thought of whether it is good thing, all done in the name of multiculturalism. Any voice of protest has to be crushed with use of words such as racism.

Not me, but many people of this country followed the Christian faith which was all part of the indigenous culture and creed of what made Britain "great". Now every other culture and religion has to take precedence over ours.

Religious slaughter is of particular interest to me, it goes against everything I have ever stood for with regard to food animal welfare. It has to be said that many animals destined for this trade are in fact electrically stunned, with which I do not have a problem. Many however are not, and a Halal abattoir not far from where I live will stun on a Monday, Tuesday, Thursday and Friday but not on a Wednesday, so as to accommodate the so-called highly religious zealots who will not accept any form of stunning. How anyone in the name of religion can deliberately inflict agony on a food animal is beyond my comprehension.

It sounds as though I am racist, this is certainly not true, nobody can help how he is born. I have friends who do not have the same colour skin as me but I call them friend and judge my relationship with them as I would with any other individual. What is important to me is what sort of a human being you are, not what colour you are or what part of the world you come from. I have, however, a problem with other races and cultures diluting our cultural heritage in this country, and I have a problem with various religions. The Pope recently said that atheists like me would be forgiven by his God if they lived a sober and decent life. How arrogant is that? How condescending of him to think that he has the right to judge me! He cannot even take action against many of his priests (forbidden sexual contact in the normal sense of the word by their religion), who are guilty of sexual abuse of children, ruining young lives. I think it is he that should be looking for forgiveness for his religion.

There are many other things in today's world that I find sad or at the very least distasteful but I am going to leave it there and thank my readers for coming this far with me and I hope they may have found my reminiscences interesting.

EPILOGUE

I HAD FINISHED this book and everything was being brought together for the final publication when an article was published in the Daily Mail written by John Stevens on the 24 February 2014. I was so incensed by what I read, that I had to reach for my pen again.

The report concerned a traditional butcher plying his trade in the market town of Sudbury in Suffolk. For at least thirty years the business had attracted customers with a shop window display of poultry, rabbits and various game still in fur or feather. Intermixed with this more unusual fare were pigs' heads. This was in an area where, probably like many other rural areas of the country, ex-town dwellers had infiltrated the rural idyll and brought their sad, sanitized, out of touch with the real natural world attitudes with them.

They turned their ignorance on this unfortunate butcher and he received hate mail and some people wrote to the local paper whilst others put their vicious remarks on something called Facebook. This is where people who have not the guts to face people on a one to one basis hide their vitriolic attacks from the safe distance of their homes.

For many generations high street butchers have advertised their wares in this way; they would compete with one another as to who could put on the best display. The high street butcher has for many years now been in horrendous decline, due to the march of the unstoppable monster of the supermarket stranglehold. A surviving small butcher has inevitably to turn to a niche market to supply the more discerning customers. These customers probably expect to know where these animals and birds had originated from; in fact in many cases traceability from source is expected. It has to be said that butchers shop displays have to be done correctly. Animals in the form of rabbits and squirrels hanging with red and white meats and game birds and poultry still in the feather, is probably not a good idea.

Consider for a moment food animals and birds which when they are ready for slaughter, are taken to abattoirs to be made into meat. You cannot eat them alive and so humane dispatch has to come into the equation somewhere, unless of course they are unlucky enough to be selected for religious barbaric slaughter as previously described. Make no mistake when you see the lambs, calves, piglets and chicks in the fields and on television none of them would exist were it not for the fact that we eat meat and are genetically evolved omnivores, not carnivores because we enjoy a mixed diet from our days as hunter gatherers. True carnivores like the wolf, the predecessor of our best friends in the animal world, when having killed a herbivorous animal will eat all of it, including the stomach contents which have had the cellulose layer of the grass or leaves that they have consumed broken down by the herbivorous animal's digestive system making the food value available to the carnivore which cannot break this down for itself. You only have to look at the faeces of a dog which has eaten grass, to see that it comes out much the same as it went in. We have all seen nature programmes on the television where a pride of lions have killed a wildebeest and the first thing they do is to open up the body cavities to make sure they all get a share of the viscera and the stomach contents to give them the trace elements that their bodies need. Interestingly various parasites have evolved over millennia to accommodate this sequence of events, and the intermediate stages of various tapeworms have positioned themselves in a herbivorous animal's viscera to be consumed by and developed into the final stage in their ultimate host, be that dog (wolf) or in two cases man. There are in fact two human tapeworms to be found in their intermediate stage; in the one case cysts in the musculature of cattle, and in the other case cysts to be found in the musculature of pigs. When teaching a group of students, I often enquire as to whether or not there are any vegetarians amongst them, there is usually a certain amount of hesitation and eventually one will say yes they are and I enquire as to why? They may say that they do not like animals to be killed for their benefit. This is a perfectly acceptable reason, another may say they do not like the thought of animals and birds being intensively farmed, I agree entirely. Another may say I do not like the taste of meat that too is OK but another may say we are not meant to eat animals and for the reasons I have described that is definitely not the case. The tapeworms I have described would not have evolved over millions of years, were it not for the fact that we eat meat as part of our diet.

People who criticise this poor butcher for displaying meat in its original form are denying our heritage and bringing up generations of youngsters who now believe that meat is made in a factory somewhere, milk somehow is put into plastic cartons and eggs appear in cardboard boxes. No! Meat is the flesh of food animals which have to be killed, milk comes from cows' udders and eggs come out of chickens' backsides. Even potatoes are the tubers of a plant which has to be dug up, not a powder or chip made in a factory somewhere.

What a sad indictment of the modern world which has become so far divorced from nature. Sadly when the crunch comes, and come it will, our youngsters will not know how to catch a rabbit for supper. Recently those poor people in Syria who are starving to death have to try and catch mice and rats to eat to save themselves from starvation. At least in some other less developed countries they know how to go about it.

The only good news to come out of this sorry ongoing saga is that people from all over the country have, like me, been so incensed by this crazy story that they have contacted the butcher concerned to express their support for him in his traditional continuation of his shop display. Butchers' displays have been going on for generations and just because some vitriolic modern day ignoramuses decided that they are offended by it, and it might upset their children, the butcher decided to remove the display.

The good news at the end of all this is that the small butcher concerned has been so heartened by the support he has received that he has decided to reinstate his displays. Perhaps the worm had turned at last and we can return this country to again being GREAT Britain.

I just hope that the local population support him and his business. My experience over fifty odd years is that children are very resilient and if death, be it human or animal is simply and carefully explained to them, they seem very able to adjust to the fact that death is part of life. Many adults who have lived in a cocooned environment all their lives seem far less able to accept this very basic concept.

* * *

On a completely different subject the results of the badger cull in Somerset have just been announced in the media. Apart from the complete failure to achieve the magic number, predicted by the so-called "experts", a far more worrying fact has come to light. A percentage, which seems to vary between five and eighteen percent, of those badgers that they managed to kill, were not killed in a humane manner and reportedly took more than five minutes to die. One can only speculate as to how this information was achieved. It begs the question as to whether there was someone standing there with a stopwatch while the killing was being done, to time the procedure from start to finish. Also, what about the ones that may have been wounded and run away, maybe down into a sett, to die a long lingering death in agony. Some may not even die but given time will recover to live the rest of their lives horrendously debilitated. It is no wonder it costs the tax payer thousands of pounds for each badger killed.

As I understand it the dead badgers were taken back to a secure site in Gloucestershire to be looked at to decide whether or not they had been humanely killed, not as one might have expected to be post-mortemed to see if they had bovine TB or not. Of course had this been done it would have had to be publicly exposed under the Freedom of Information Act. It easily could have revealed that a tiny percentage were actually suffering from the disease which would not have looked good as far as the case for culling of thousands of badgers is concerned. I would take my reader back to the colony near Stroud in in Gloucestershire many years ago when everything pointed to the justification of gassing a large badger colony. When later excavated and sixteen dead badgers were dug out and subsequently post-mortemed, none were found to have bovine TB. The politicians have learnt a thing or two since then and found to their advantage that if you do not actually look for something you are never going to find it. As with so many other things today this was for political expediency and you must never forget votes are everything; decency and honesty must be kept to a minimum. For the record, while working for MAFF as I did during the horrendous days of the foot and mouth period, I know that if it had taken my partner or me more than one shot to kill a beast or a sheep or a pig in those frenzied days we would have been thrown off the job, and rightly so. It seems today that you can watch a badger taking five minutes to die, what is wrong with shooting it again or just possibly have sufficient expertise for it not to have happened in the first place. It doesn't matter whether it is five percent or eighteen percent one animal made to suffer in this way is one too many.

* * *

I really thought I had covered more or less everything I set out to cover. Then in March of 2014 another bombshell appeared, when John

Blackwell a very brave and highly principled vet who is the recently appointed chief of the British Veterinary Association pronounced that religious slaughter of food animals without stunning should be banned, because it **is** cruel. Other countries in Europe have already banned it, something they are able to do under EU law.

This has caused uproar, particularly amongst the Jewish community and to a lesser extent the Muslim community. Remember from my chapter on religious slaughter that none of the animals slaughtered under the Jewish method for Kosher meat are stunned, whether the Jewish slaughterman has been well trained or not. Muslim slaughter has come to accept that many of the animals which are killed for Halal meat are in fact stunned but must have their throats cut by a Muslim. As I became aware many years ago at Gordon Road in Bristol there are however different levels of religious belief and the higher levels will, as with the Jews not accept any stunning.

The Jewish communities with many individuals highly placed in both government and business circles have come out and said that their form of barbaric killing is **not** cruel. You only have to ask yourself if you found yourself in a position in which you were going to die by having your throat cut, if you were given the option of just having that procedure performed, or being rendered insensible to pain prior to having your throat cut, which would you choose? I think the answer would be clear.

Of course the other aspect of which I think most people are unaware is the fact that far more animals are killed in this barbaric way than are used by either of these two religions. Inevitably much of this tortured flesh ends up in the general meat trade and is consumed by the non Muslim or Jewish population. Added to the horseflesh scandal and the tubercular flesh scandal now the general public finds out that they are eating flesh derived from animals which died in agony.

When the news broke, first of all that deceitful twerp Clegg announced that he would not support a ban which compromised religious rights, and now while in Israel on a visit (how appropriate) the other two faced twerp Cameron announces yet again that, there will never be a ban on his watch. Obviously, there is too much Jewish pressure amongst their party members as well as amongst influential big business.

Full marks to John Blackwell, perhaps he and others like him with a sufficient sense of decency and plenty of backbone could run our country and maybe bring it back to the once world-wide envied state of "GREAT".